Seismic Damage Assessment Of Reinforced Concrete Structures

Low, Medium, And High Rises Using A Performance-based Seismic Design Approach

Amey R. Khedikar

Title: **Seismic Damage Assessment Of Reinforced Concrete Structures**

Low, Medium, And High Rises Using A Performance-based Seismic Design Approach

ISBN: 979-8-89966-407-6

Author: Amey R. Khedikar

Cover image: AI Generated

Publisher: Generis Publishing
Online orders: www.generis-publishing.com
Contact email: info@generis-publishing.com

CONTENTS

TABLE INDEX

FIGURE INDEX

ABSTRACT

Performance-based seismic design provides a methodology of understanding the risk to life, occupancy interruption and economic loss that may occur as a result of seismic events. The methodology has two primary concerns; appropriately quantifying the uncertainties associated with the performance evaluation process, and satisfactorily characterizing damage sustained by the structure. Structural engineers have been using nonlinear static and dynamic analysis, from which nonlinear static analysis (pushover) became common, due to its simplicity. In pushover analysis, the structure is subjected to lateral load that represents approximately the relative inertia forces generated at locations of substantial masses such as floor levels. The static load pattern is increased in steps and the lateral load-roof displacement response of structure is determined until a specific target displacement level or collapse is reached. Even if the buildings were designed according to the conventional earthquake-resistant design philosophy (force-based approach), earthquake disasters have shown that significant damage occurred even if the buildings were designed according to the conventional earthquake-resistant design philosophy (force-based approach). An alternative to force-based processes was determined to be the performance-based seismic design framework's nonlinear static assessment methodologies. Preliminary estimations of prospective damage to tall buildings have become important because of the rising demand for such structures, in order to reduce repair costs and assure public safety. An illustration of the plastic collapse process can be shown using nonlinear evaluation tools, however the damage to buildings can't be accurately scaled. In this work, a set of damage indices were developed to measure the extent of structural damage. The damage indices were constructed using engineering demand parameters gleaned via nonlinear analysis.

The suggested damage indices were found to be a reasonable way to scale the damage value in the examination of the example MRFs. In this work, performance-based design is shown to be particularly beneficial in R.C. structures of all heights (Low Rise, Medium Rise and High Rise), from the ground to the sky. When it comes to seismic load bearing capability, Performance-Based Seismic Design creates buildings that are more resilient to earthquakes while also being safer and more cost-effective. In addition, the PBSD created via this approach meets the acceptance requirements for immediate occupancy and life safety limit states for a variety of seismic intensities.

CHAPTER 1.
INTRODUCTION

1.1 General

An earthquake's ground shaking might be minimal, moderate, or violent depending on the area. Minor tremors are common, moderate tremors occur quite seldom, and violent tremors are extremely unusual.

Engineers are now working to create earthquake-proof structures that will not be harmed even in the event of a powerful earthquake, which is unlikely but possible. Rather, the sole engineering goal up until the last decade has been to design earthquake-resistant structures, which can withstand ground shaking even if they are badly damaged but would not collapse in the event of a large earthquake. In earthquake-resistant structures, the protection of people and their possessions is ensured, and therefore a calamity is averted. Earthquake safety is the primary goal of all seismic design regulations across the globe.

From a historical perspective, it's easy to observe that seismic structural design has evolved far more rapidly than other types of load instances and activities, such as gravity, wind, and traffic. First, after the major earthquakes in the early twentieth century (Kanto, Long Beach and Napier), seismic assault was seen as a basic mass-property lateral force, with elastic structure action resisting it. However, structural analysis was still carried out on the basis of elastic structural response, which meant that structural period was not yet integrated into structural design during this time period. The practical and empirical findings that well-detailed structures could withstand ground shaking levels capable of creating inertia forces several times greater than those anticipated by elastic analysis led to the introduction of ductility concerns in the 1960s and 1970s. Considerations of ultimate strength were based on force levels decreased from the elastic values by what seemed to be arbitrarily assigned force-reduction factor values that vary dramatically across the design codes of various seismically active nations. Increasingly, individuals are realizing the importance of strength in helping to prevent displacement or strain, both of which have been connected to damage potential. Therefore, deformations, not strengths, should be used to define structural vulnerability.

1.2 Overview

It has become necessary to understand the effects of seismic forces before planning reinforced concrete structure in earthquake prone areas. Earthquakes may be induced naturally or from man-made sources. Volcanic activities, geological fault rupture and landslides are the natural sources of earthquakes. Mine explosion, nuclear testing, etc. is man-made sources of earthquakes. Earthquake may be defined as a sudden release of strain energy in the form of elastic waves in the Earth crust. Ground shaking produced by earthquakes might be minimal, moderate or violent in nature. It depends on the dissipation of strain energy by rocks. Minor tremors are common, but don't have any associated losses. Moderate tremors occur quite seldom and violent tremors are extremely unusual, but results in significant loss of property and life [Ghorbah A., 2001; Murthy et al., 2012].

Reinforced Concrete (RC) structure when subjected to seismic hazards (earthquakes) results in inelastic excursions. Force-based design (FBD) methodology documented in present seismic codes provides a rational approach to address this inelastic behaviour by the use of response modification factors. These response modification factors are used for strength and displacement amplifications. Structures produced from these gravity-based design approaches were found to be get damage and or collapsed during a seismic event, thus questioning the adequacy of the FBD design philosophy [Zou and Chan,2005; Mondal et al, 2013].

Performance-based seismic design (PBSD) has emerged as a better alternative to the FBD [Hamberger R., 2005; Ghorbah A., 2001; Zameeruddin and Sangle, 2016]. PBSD is an iterative process wherein structures are designed for pre-defined structural and non-structural performance objectives. These performance objectives are statement of acceptable risk in terms of repair, downtime and casualties resulting from the damage state of structural and non-structural components of a building. Performance levels are defined in terms of drift limits. Preliminary designs are prepared for the selected performance objectives. Followed by the performance evaluation of the structure, for which the capacity of the structure is compared with the demand imposed on it (FEMA 440).

PBSD in its present state is capable to evaluate the capacity of the structure, but fails to provide any damage value. The present study aims to provide a rational approach to identify a damage level or scale a damage value using the variables resulting from the performance evaluation method. To do so a series of low-rise, medium-rise and high-rise example moment resisting frames, design as per FBD were subjected to seismic loads Nonlinear static approaches based on pushover analysis are used in PBSD publications for performance-based seismic assessment. Pushover analysis (POA) is utilized for an approximation analysis, POA employs a sequence of progressively elastic experiments piled on top of one another to approximate the overall structure's force-displacement curve. (Zameeruddin and Sangle, 2016; Ghobarah, 2000).

An initial gravity load is delivered to a two- or three-dimensional model that contains all lateral force resisting components in bilinear or trilinear load-deformation diagrams. Afterwards, a predetermined pattern of lateral load distribution is applied along the building's height. The lateral pressures are raised to the point where some of the participants give way. POA's collapse process explains the structure's inelastic nature (Ghobarah, 2000).

The current study employs a logical method to determine the performance of a gravity-based design RC structure under varied lateral loads. SAP 2000 V 17.0, a soft computing tool, was used to perform POA based on displacement. (2000) Wilson and Habibullah Engineering demand metrics such as displacement, base shear, storey drift, inter-storey drift, and response modification factor have been used to verify a variety of acceptable performance-limit states. A performance indicator is also used to analyze the structure's performance. In order to check for consistency across time history study, the structure was designed with two alternative patterns of plastic hinge position.

1.3 The need for a seismic design based on performance (PBSD)

Bureau of Indian Standards-based rules are often used to control design and construction in India's states and municipalities. To ensure the protection of life and property from fires and other risks, building regulations must be enacted and implemented by local authorities (IS 456, 2000).

Prescriptive standards, such as those governing permissible building materials, permitted structural and nonstructural systems, needed minimum levels of strength and stiffness, and control over the specifics of how a structure is to be put together, are used to achieve this objective. Based on a wide range of categories for structures and occupants, prescriptive criteria sometimes include fixed values such as fire resistance ratings and permissible area or height as well as standards linked to structural design (IS 875, 1987). Traditional code design does not take into account the actual performance capabilities of specific building designs, despite the prescriptive criteria used in model building codes being designed to produce structures capable of meeting specified performance standards. With prescriptive design, the performance capabilities of a particular building is not always understood for a given structure. Buildings created to these prescriptive requirements may perform better or worse than the code's minimal expectations, depending on the specifics of the project (FEMA 445,

2006).

Improvements in the development of earthquake design standards are continuing. If you look at earthquakes in India and throughout the globe, you can see how the Bureau of Indian Standards (BIS) seismic design rules have evolved over time. Regulations requiring a minimum degree of lateral strength were developed in response to the earthquakes indicated in Table 1.1 throughout the twentieth century. It was important for structures to have the capability of deforming without causing catastrophic collapse, which is called ductility.

Table 1.1: Earthquake-related deaths and property loss in India

(Zameeruddin and Sangle, 2017b)

Sr. No.	Earthquake	Year	Intensity	Fatalities
1	Latur (Khillari)	30, September 1993	6.2	9,748
2	Chamoli	29, March 1999	6.8	103
3	Gujrat (Bhuj)	26, January 2001	7.6/7.7	20,000
4	Off west coast (northern Sumatra)	26, December 2004	9.1	15,000
5	Kashmir	8, October 2005	7.6	1,350
6	Gangtok (Sikkim)	18, September 2011	6.9	118
7	North India	25, April 2015	7.8	8900
8	North India	12, May 2015	7.3	04
9	Dibrugarh (Assam)	28, June 2015	5.6	3 injured

This belief is shared by most property managers as well as most design experts, who don't typically take into account predicted levels of damage when designing a new structure or remodeling an existing one. Buildings developed in compliance with the current viewpoint design rules have been required to recognize that damage, sometimes severe, might occur in earthquake-prone structures.

Engineering practice and research have changed as a result of advancements in seismic design criteria and changes to the state of knowledge. Seismic performance assessment is moving toward more predictive methodologies in order to provide stakeholders the information they need to make reasonable economic or safety choices. Meanwhile, academics have been developing new analytical tools and data sets that will help enhance evaluation processes. The development of performance-based engineering methodologies for seismic design has resulted from the recognition that code-based strength and ductility criteria are not adequate for the assessment and upgrading of existing structures.

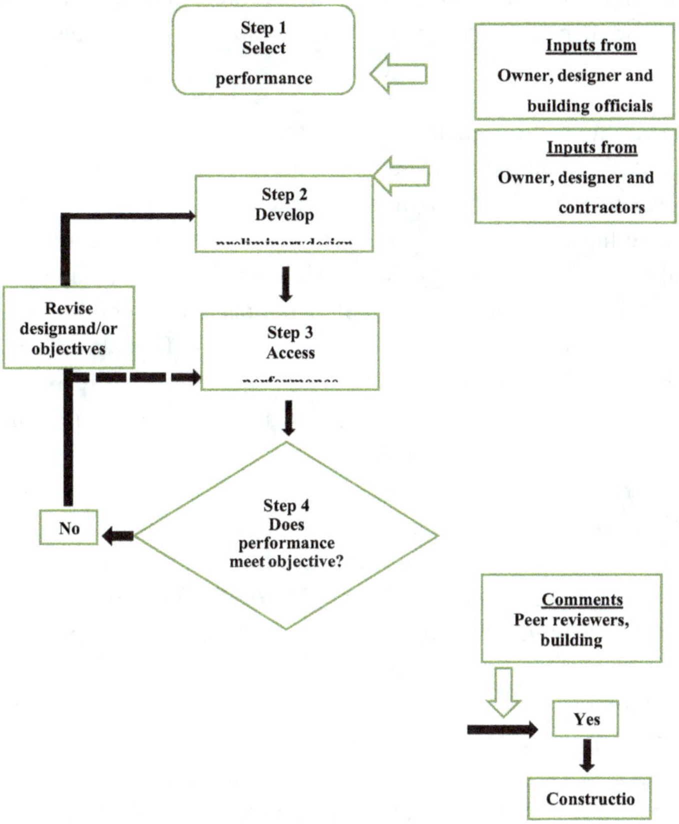

Figure 1.1: ***Schematic of the flow for performance-based seismic design***

Design requirements are described in terms of meeting given performance goals when the structure is exposed to the indicated levels of seismic danger, known as Performance-Based Seismic Design (PBSD). The American Society of Civil Engineers (ASCE) created PBSD in the 1970s. Buildings may be designed and built using the Performance-Based Seismic Design (PBSD) standard with confidence, knowing exactly how vulnerable they are to future earthquakes and how much damage they will do to people inside and the economy as a whole. To get to the desired performance level, the PBSD process involves a series of iterative steps that begin with the selection of performance objectives (defined by owners, designers, and building officials), continue with the creation and evaluations of preliminary designs (based on stated performance objectives), and conclude with a series of revisions and reevaluations as necessary. There is a framework provided by this method for determining the levels of security and property protection required for a project and for determining expenses acceptable to the owner, the designer, and local building authorities depending on the specific project requirements.

The creation of a performance-based engineering (PBE) approach, in addition to the method's further simplicity, is the focus of the efforts of a number of scholars. It was suggested by Huang, Y. N., Whittaker, A. S., and Hamburger, R. O. that a simpler technique using linear analysis might be used for low to medium-rise structures when employing the PBSD methodology (2017). Hajirasouliha, I., Asadi, P., and Pilakoutas, K. came up with a practical and simpler approach for reinforced concrete (RC) constructions based on the notion of distribution for deformation. This method was suggested by the three researchers (2012). The approach recommends moving materials from areas that are strong to parts that are weak in order to achieve consistent deformation. It was proved that a PBSD technique may be used for RC structures with infill walls, with life-cycle cost serving as the performance measure. In the framework of structural optimization by means of an evolutionary algorithm, a comparison was made between the method that is predicated on non-linear static and dynamic assessments. Cardone, D., Palermo, G., and Dolce, M. came up with an idea for a draft code that would apply to buildings in Taiwan and use a PBSD approach (2010). The goals of the seismic design were chosen based on the various user groups, and the qualitative interpretation of performance requirements, which included drift limitations, was used. In the field of performance-based earthquake engineering, a number of academics are further developing a technique that is based on displacement. Several researchers, including us, have proved the feasibility of displacement-based seismic design inside the PBSD domain. The authors Cardone, D., Palermo, G., and Dolce, M. came up with the idea of using a direct displacement-based design (DDBD) technique for seismically isolated structures (2010). They performed a nonlinear time history analysis (NTHA) on base-isolated structures with a variety of layouts, and based on their findings, they made several suggestions for improving the DDBD technique. Yield Frequency Spectra (YSF) was implemented for the preliminary performance-based seismic design with a set of performance objectives by Vamvatsikos, D., Kazantzi, A. K., & Aschheim, M. A. (2015) & Vamvatsikos, D., & Aschheim, M. A. (2015). Yield Frequency Spectra (YSF) was implemented for the preliminary performance-based seismic design with a set of (2016). The YFS has been represented as the mean annual frequency (MAF) that is greater than the threshold value of ductility or displacement. Franchin, P., Petrini, F., and Mollaioli, F. developed a probabilistic, risk-targeted process using annual frequency (MAF) of exceedance as performance targets. This was done in order to achieve the best possible results (2018). This technique enables direct management of the seismic risk posed by the structure in numerous limit states, fulfilling both the structural design variables and the capacity design criteria.

Although there have been attempts made by certain researchers to take into account the soil-structure-interaction (SSI) impact when designing for seismic activity, these attempts have not yet been included into the methodology of performance-based seismic design principles. In the construction of buildings to withstand seismic hazards, the impacts of SSI were disregarded since it was assumed that it would have positive consequences. Foundations that are flexible may have an effect on how buildings respond when placed under seismic stresses, and the benefits of using SSI are discussed in the most recent standards for seismic design published by the American Society of Civil Engineers (2016). The seismic response of structures that are susceptible to seismic forces is significantly influenced by the SSI, which also has an effect on the dynamic characteristics of the system. Fatahi, B., Hokmabadi, A. S., & Samali, B. Fatahi, B., Hokmabadi, A. S., and Samali, B. conducted an investigation using a shaking table test to determine the effect that different kinds of pile foundations have on the seismic performance of buildings. The rocking component is primarily responsible for the changes that may occur to the dynamic characteristics of the system. These changes can result in an increase in lateral deflection and inter-story drifts, both of which have the potential to affect the performance level of the structure. Because performance-based seismic design means an increased emphasis on displacements rather than on inertia forces, adequate evaluation of SSI as a significant element is essential. According to the findings of several studies, the interaction between the soil and the structure may also have a negative impact on the structural response. Both Mylonakis and Gazetas (2000) as well as Priestley, M.J.N., and Park, R. (1987). As a result, it is understood that the impacts of SSI do not necessarily have favorable benefits, and they may even have effects that are very negative for the response of the superstructure.

1.4 Benefits of using a performance-based approach for earthquake design

In opposition to the prescriptive design techniques, performance-based design offers a methodical process for evaluating the capacity of performance possessed by a building, system, or component. This is done in contrast to the prescriptive design approaches. It is possible to utilize it to test the equal performance of alternative solutions, to give standard performance at a lower cost, or to validate better performance that is required for essential facilities.

In addition to this, it offers a language that makes it easier for stakeholders and design experts to have meaningful conversations about the creation of design possibilities and the choices to be made among them. Based on the unique demands of the project, it gives a framework for assessing how much safety and how much property protection are acceptable to stakeholders at what expense. This can be done by looking at the level of protection needed for the property and the level of protection needed for the property. Design for earthquakes based on performance may be used to; (Zameeruddin and Sangle, 2016; FEMA 445, 2006)

1. Individual structures should be designed with a greater degree of confidence that the performance that is anticipated by the current building regulations will be realized.

2. Develop individual structures that are capable of achieving the performance requirements of the existing building regulations, but that do so at a reduced overall cost to construct.

3. When designing individual buildings, be sure to design them to attain greater performance (and fewer potential losses) than what is anticipated by the current building regulations.

4. Design individual structures that go above the boundaries given by the code in terms of their configuration, the materials they use, and the systems they implement in order to achieve the level of performance that is intended by the current building regulations.

5. Determine the current structures' potential seismic performance and estimate the possible losses in the case of a seismic event.

6. Conduct an analysis of the potential performance of the current prescriptive code requirements for new buildings, and use the results of this analysis as the basis for making improvements to the seismic design criteria established by the code. This will allow future buildings to perform more consistently and reliably.

The implementation of a performance-based seismic design has the potential to make the allocation of societal financial resources for the purpose of mitigating the effects of future earthquakes more both efficient and effective. In addition, the technology that is utilized to execute performance-based seismic design is transferable, and it is possible to be adapted for use in performance-based design for other severe hazards such as fire, wind, flood, snow, explosion, and terrorist attack (ATC 58, 2005).

1.5 Performance-based processes of the first generation

The formal approach of performance-based design was created in response to the seismic design dilemma that developed in the 1990s. As a result, it was impossible to reliably or feasibly apply the code-based strength and ductility standards for new buildings to older structures.

In 1992, FEMA began developing the first set of performance-based seismic design procedures as the culmination of its efforts to reduce the seismic hazards provided by existing buildings. Prior to the FEMA 273 Report NEHRP Guidelines for Buildings Seismic Rehabilitation (ATC 40, 1997) and its companion document the FEMA 274 Comments on the Guidelines for Buildings Seismic Rehabilitation (ATC 40, 1997), which dealt with seismic upgrades for existing buildings, this preliminary work was completed. The Structural Engineers Association of California also published the Vision 2000 Report, Performance-Based Seismic Engineering of Structures concurrently (SEAOC, 1995). Performance-based seismic design guidelines for new building construction were presented in this paper. They introduced the basic concepts of performance levels in relation to damageability and various levels of risk (Freeman, 2004; ATC 40, 1996 and FEMA 273, 1996).

Initial procedures defined performance in terms of a set of distinct performance levels, each with a name indicating the projected degree of damage: collapse, life safety, immediate occupancy, and operational performance. Their work also revealed performance as it related to structural and nonstructural component deterioration. We were able to develop some performance goals by connecting each of these levels of performance to a certain degree of seismic hazard.

For the first-generation methods, nonlinear analysis methodologies and acceptance criteria, as well as a collection of analytical processes that could be utilized to model the seismic response of structures ranging in difficulty, were supplied. Earthquake simulations were conducted using these methods. Compared to the current building code processes, these first-generation procedures were a considerable breakthrough since they gave a logical approach to the design of buildings in order to achieve a specified performance level (FEMA 445, 2006).

1.6 Performance-based processes of the second generation

At the moment, performance-based seismic design approach is usually based on execution of processes and criteria provided within the FEMA 356 Prestandard and Commentary for the Seismic Rehabilitation of Buildings. This standard was developed for seismic retrofitting of existing buildings (ASCE, 2000).

With the release of FEMA 356, an enhancement to the first-generation methodology specified in FEMA 273 has been made. the FEMA 343 Report, Case Studies: and the use of these techniques in engineering practice Evaluation of the NEHRP Guidelines for the Seismic Rehabilitation of Buildings, was utilized in developing FEMA 356 to incorporate new technical standards and criteria for FEMA 273. The knowledge gathered from the usage of the processes in engineering practice prompted these revisions (FEMA 445, 2006).

Engineering practitioners have a better understanding of the principles behind performance-based procedures as a result of the creation of second-generation performance-based procedures. Both performance-based seismic design and the application of sophisticated nonlinear analytic methods are becoming more commonplace in engineering practice. Performance-based seismic design has become an industry standard. Despite the fact that the methods were developed with existing structures in mind, they are now being expanded for use in the performance-based design of brand-new buildings.

Due to the present second generation's greater usage of performance-based processes, a knowledge base of practical experience on designing with performance-based criteria and interacting with stakeholders on performance-based design challenges has been built, which has led to the development of the second generation (FEMA 445, 2006). By enhancing the coefficient to take into account soil interaction and nonlinear responses, a process of the next-generation has given a considerable improvement over the techniques of the first and second generations. They contain hints pertaining to the (Maffei, 2000)

1. Make the essential definitions and techniques of performance-based seismic design clearer and more straightforward, in accordance with the principles of capacity design.

2. Relate, contrast, and simplify procedures that are based on force and methods that are based on displacement in order to determine whether or not a structure is acceptable according to seismic standards.

3. Take into consideration whether or not it is necessary to use nonlinear analysis methods.

4. Establish defined criteria for the performance-based design of various kinds

of structures and varying degrees of their performance.

1.7 PBSD procedure:

In most cases, the selection of performance targets for a building will include the participation of a group of decision makers. This group will typically include the owner of the facility, design experts, and building authorities.

After the performance goals have been established, a number of simulations (studies of the building's reaction to loads) are carried out in order to assess the likely performance of the structure under a number of different design scenario events.

The design is considered to be complete if the simulated performance either meets or exceeds the performance targets; otherwise, it has to be revised. The PBSD technique is represented in the form of a flowchart, which may be found shown in Figure 1.

In general, for the purpose of studying Nonlinear static method (pushover analysis, capacity spectrum approach) and nonlinear dynamic process are utilized based on the needs of the structure. Linear static procedure (time history analysis).

Due to the insufficiency of elastic studies in accurately predicting the force and deformation distributions after damage has begun in a building, pushover analysis is best suited for performance-based seismic design. This makes pushover analysis the ideal choice for the performance-based seismic design. For the purpose of determining the mechanisms of failure and the extent of the potential for progressive collapse, inelastic analytical approaches are required.

1.8 Performance-Based Design: An Analysis and Evaluation

Current seismic design is performed with code prescriptive Forced Based Design (FBD) approach. In the current approach, design professionals intend to satisfy the code by selecting and designing components. Although the prescriptive criteria are intended to provide performance level of life safety, the actual performance of the structure is not assessed actually. During a design level earthquake, a code-designed building could provide strength and ductility for life safety, but could experience extensive damage leading to enormous economic loss or demolition. Extensive damages to structures designed with current codes, huge economic loss, high repair cost, etc. due to previous earthquakes (e.g. the 1994 Northridge and 1995 Kobe earthquake) imply that Force Based Design approach has some deficiencies. Thus, the decision makers, owners, and design professionals become worried that the performance of structures even designed with current codes. It has been realized that safety would not be enhanced with increased strength and could not reduce damage. This realization triggers the development of Performance-Based Earthquake Engineering (PBEE). Researchers have also found that structures designed with the PBSD approach showed better performance than the code- designed.

The frame of view is what fundamentally differentiates the process of designing brand-new buildings from the process of evaluating already-existing structures. When it comes to design, the goal is to produce a new structure that is capable of withstanding the anticipated forces (both horizontal and vertical) while maintaining an adequate margin of safety. The needed sections of the structural components of the structure need to be established for the material that will be used, beginning with a structural model of the building and the anticipated applied forces. It is standard practice to choose a model that is somewhat more conservative, which means to disregard the positive effect of certain factors. This is done for two reasons: first, to make the model more straightforward, and second, to err on the side of caution. In addition, the strength of the material is often multiplied by a certain strength reduction factor. On the other hand, the predicted applied forces are increased to account for the fact that there are unknowns.

The selection of the strength reduction factors and the design forces is governed by the goal of economic optimization; however, they are typically selected to keep the risk of damage extremely low. For example, in the design of buildings, this compares to an accepted annual probability of about 0.01 percent for achieving the ultimate capacity. In earthquake engineering, a logical design takes on more importance, and the engineer must be willing to tolerate a larger level of risk.

In this case, the yearly likelihood of reaching the maximum capacity might be as high as one to three percent depending on the situation. The purpose of an assessment is to determine how a preexisting structure will react when subjected to a set of prescribed forces. This is analogous to a study of a building structure, in which details such as the structural components, the materials, and the dead loads are provided. It is not the goal of this exercise to calculate the worst-case scenario by picking a conservative model and making conservative assumptions about the properties of the material. Rather, the goal is to evaluate the behavior that is most likely to occur when the building is subjected to the applied action. Therefore, the actual material qualities and the actual loads should be used in the analysis, rather than any safety considerations, since the latter would cause the findings to be inaccurate. In addition to this, the model should be accurate to the greatest extent feasible by taking into consideration all of the structural aspects that contribute to the process of bearing the applied forces.

In earthquake assessment programs, where the potential for damage in a particular region is calculated in order to determine effective solutions for risk reduction, the appraisal of existing structures plays a significant role.

1.9 Engineering of earthquakes based on their performance is currently being developed

Existing seismic loading regulations in building codes concentrate on the minimum lateral seismic pressures for which the structure must be constructed. However, these provisions do not clearly reflect the demand and response characteristics of the building. However, the requirements of the lateral forces alone are not sufficient to achieve the required degree of protection in a structure when it is exposed to anticipated earthquakes of varying intensity.

Based on past experiences, it has been my observation that once a methodology and its associated processes have been included into a standard or code, it is very difficult and time consuming to make any modifications. Before any of the so-called simplified techniques for Performance-Based Seismic Design and their accompanying methodologies are included into a building code, it is generally agreed that they should first undergo extensive calibration. The difficulty that arises therefore is, against what should one calibrate? There is no question that it would be great to calibrate them using the most advanced and reliable technique and process for Performance-Based Seismic Design that can be established, since this would be the perfect situation. Around the year 1990, individuals working in the field of design all around the world started taking an interest in the creation of performance-based design ideas. In contrast to the current building code provisions, which are of a prescriptive nature and mandate that structures be designed with a minimum specified level of strength and stiffness, performance-based procedures enable the designer to directly demonstrate that a design is capable of meeting certain standard performance objectives. This is achieved regardless of whether or not the design satisfies prescriptive levels of strength and stiffness.

Guidelines on how to apply the idea of PBSD to new and existing constructions can be found in documents published by SEAOC (1995, 1996, 1999) (Structural Engineers Association of California, USA), ATC-40 (1996) (Applied Technology Council, USA), FEMA-273, 274 (1997), FEMA-356 (2000), and FEMA-350 (2000) (Federal Emergency Management Agency, USA). Documents such as these explain how to put PBSD to use in practice. The 'NEHRP Recommended Provisions for Seismic Regulations for Buildings and Other Structures,' as well as more recent codes such as ICC (2000 and 2003) (International Code Council) and NFPA 5000 (2003) (National Fire Protection Association, United States), allow for the use of Performance-Based Seismic Design.

PBSD and performance-based seismic design (PBSD) are better understood after reading the aforementioned articles and books. According to ICC (2000), several recommendations and the International Code Council (ICC) (2000) have previously offered explicit quantification of different Performance-Based Seismic Design Objectives (PBSDO).

SEAOC Vision 2000 (Structural Engineers Association of California, USA). The Vision 2000 Committee of the Structural Engineers Association of California (SEAOC) published a report in 1995 titled "Performance-Based Seismic Engineering of Buildings," which outlined a promising approach toward the aforementioned development. This approach will be referred to as "Performance-Based Seismic Engineering" (PBSE), but it is also known as "Performance-Based Earthquake Engineering." In 1995, the PBSE was proposed by the Vision 2000 Committee of the SEA (PBEE). The report referenced above presents a conceptual framework for PBSE, in addition to the various methodologies that have been proposed for the application of such framework to the design, construction, occupancy, and maintenance of the building in question, with a particular emphasis on the design that has been dubbed "Performance-Based Seismic Design" (PBSD).

1.10 Philosophy of Design for Earthquakes

To summarize the earthquake design concept, below are a few key points.

The primary structural components of the building, which carry vertical and horizontal loads, should be unaffected by mild but frequent tremors, although non-load-bearing portions of the structure may experience damage that has to be repaired. Main members of the building may incur repairable damage, while other portions of the structure may be damaged beyond fix and need replacement after the earthquake.

When the ground shakes violently but seldom, the primary structural components may be damaged severely (and in some cases irreparably), yet the structure should remain intact. Consequently, after just a few moderate tremors, the building may resume normal operations quickly and at cheap expense. Upon completion of the repairs and strengthening of the damaged major components, the building will be ready for use after just a little shaking. Although the structure may no longer be useful after a big earthquake, it will nonetheless remain so that people and their belongings may be evacuated and retrieved.

The design of every system must take into account the possibility of harm. It is of the utmost importance that essential post-earthquake infrastructure, such as hospitals and fire stations, remain operating as quickly as is humanly feasible following the catastrophe. These structures need to be built in a way that allows them to endure a significant amount of earthquake damage, since it is one of their primary requirements. In the aftermath of an earthquake, dams can collapse, which can cause floods downstream, which in turn might start off another chain of events. As a consequence of this, dams (as well as nuclear power stations) must to be constructed such that they can resist earthquakes of an even higher magnitude. The reaction of the building to an earthquake is seen in figure 1.2.

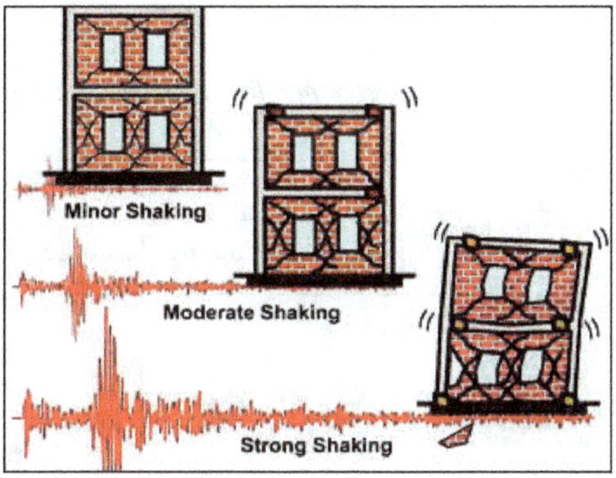

Figure 1.2: ***Performance of the Building (Murthy 2002a)***

Quakes are a natural occurrence and are caused by the constant movement of the earth. Even with all of the developments in science and technology, an earthquake cannot be stopped. People need to get used to accepting this as their lot in life. An earthquake is the kind of natural catastrophe that may be expected to endure the shortest amount of time and for which the feasible reaction time is the shortest. In the event that an earthquake is about to strike, everyone should immediately exit their homes, even if they have just ten seconds to do so.

Poorly built structures, which either completely or partly fell as a result of the earthquake, were responsible for the deaths and injuries of a large number of people. Walls crumbling amid the confines of tight alleyways, burying everyone trying to escape within them. Roofs and cantilevers that were not properly secured fell on top of individuals.

Violent shaking is bringing free-standing walls, parapets, and balconies down to the ground. Buildings with vast open areas at ground or first-floor level, such as garages or shops, collapse and bury people in contemporary reinforced concrete structures (also known as soft storey collapses). Being unable to react correctly as the walls surrounding them start to shake and fall down.

1.11 Analysis Employing a PBSD Methodology

For performance assessment following the PBSD approach, at first the performance objectives should be selected. Performance objectives are associated with a building's ability to withstand a certain magnitude of earthquake damage. For the present study, hazard levels and performance levels of the FEMA 356 have been selected which is represented in Table 2.

Table 1.2: Hazard Level and Corresponding Performance Level as Per FEMA 356

Hazard Levels \ Performance Levels	Operational	Immediate Occupancy	Life Safety	Collapse Prevention
50% in 50 Years	O			
20% in 50 Years	Δ	O		
10% in 50 Years (DBE)	◊	Δ	Basic Safety Objective O	
2% in 50 Years (MCE)		◊	Δ	O

As the structure is a residential building, the basic safety objective has been selected for performance assessment. Acceleration response spectrum for those hazard levels have been prepared. performance level of structure for the seismic hazard level that has been specified. matched with the response spectrum corresponding to the required hazard level. Nonlinear Time History Analyses (NTHA) have been performed using the matched acceleration records.

1.12 Methods of Analysis (Various Approaches to Research)

A structural study of the mathematical model of the structure is needed to estimate the forces and displacements necessary in different parts of the structure for a seismic performance assessment. Seismic performance may be predicted using a variety of approaches, both elastic and inelastic.

1.13 Elastic Methods of Analysis

An elastic analysis is used to determine the force demand on each component of the structure and compare it to the available capacity. Code static lateral force procedure, code dynamic procedure, and elastic process employing demand-capacity ratios are all examples of elastic analysis methodologies. As the name implies, these techniques rely on the assumption that buildings are elastic in their response to earthquakes.

Using the code static lateral force process, the structure is subjected to forces derived by scaling down the smoothened soil-dependent elastic response spectrum by a structural system based on the "R" force reduction factor. If the structure's actual strength is greater than the design's, it is expected that the structure may release energy by yielding.

An elastic dynamic analysis is used in the code dynamic technique, and its results are used to estimate the force demands placed on the different components. It's possible that the dynamic analysis will be a response spectrum analysis instead of an elastic time history analysis. For the purpose of response spectrum analysis, a sufficient number of modes must be taken into consideration in order to have a mass involvement of at least 90 percent. The time history analysis will always take into account any influence that is caused by higher modes.

In the Demand and Capacity Ratio method, also known as the DCR technique, the force actions and their related capacities are compared to one another as demand and capacity ratios. Calculations for DCR requirements are required to take gravity into account. The DCR technique takes the complete earthquake demand without any reduction and adds it to the gravity demands. This is in contrast to the code static lateral force and code dynamic processes, both of which decrease the full earthquake demand by an R-factor. DCRs that are getting close to 1.0 (or even higher) can suggest that there are probable issues.

Despite the fact that force-based processes are widely used in the engineering industry and simple to put into practice, they do have a few limitations. In the elastic range of strength and deformation, structural components are tested to determine whether or not they are serviceable. An elastic study is not sufficient to determine the post-elastic behavior of structures. Nevertheless, one must take into account the post-elastic behavior since, in the event of a powerful earthquake, practically all buildings are likely to undergo deformations that fall within the inelastic range. The seismic force reduction factor, denoted by "R," is used in order to indirectly take into consideration the effects of inelastic behavior by converting elastic forces into inelastic ones.

There are several limitations to the use of force-based processes, even though they are well-known in the engineering community and straightforward to implement.

The elastic range of strength and deformation is used to assess the serviceability of structural components. Post-elastic structures could not be recognized by elastic analysis. After an earthquake, practically all buildings will deform in an inelastic range, hence post-elastic behavior must be taken into account. By decreasing elastic forces to inelastic, the seismic force reduction factor "R" accounts for inelastic behavior indirectly.

The researchers may have been motivated to create displacement-based approaches for seismic performance assessment as a response to the disadvantages associated with force-based procedures as well as the reliance of damage on deformation. The displacement-based techniques employ the nonlinear analytical processes to take into consideration both the seismic demands and the available capacities explicitly. These procedures are primarily based on inelastic deformations rather than elastic forces.

1.14 Inelastic Methods of Analysis

In the event of a powerful earthquake, structures experience severe plastic deformation, and the dynamic properties of the structure change throughout the course of its lifetime. Therefore, the evaluation of the performance of a structure necessitates the use of inelastic analytical approaches that take into consideration these characteristics. The real behavior of structures may be better understood with the use of inelastic analytical processes, which reveal failure modes and the possibility for gradual collapse of the structure. The processes for inelastic analysis primarily consist of inelastic time history analysis and inelastic static analysis, which is also known as pushover analysis.

The most reliable approach for predicting the force and deformation demands at different components of the structure is the inelastic time history analysis. However, the application of inelastic time history analysis is restricted due to the fact that the dynamic response is very sensitive to modeling and the features of ground motion. A correct simulation of the cyclic load deformation characteristics, taking into consideration the deteriorating features of all-important components, is required for this. In addition to this, it is necessary to have access to a collection of representative ground motion recordings that can take into account the uncertainties as well as variances in the characteristics of severity, frequency, and duration. In addition, the amount of time needed for computing, the amount of time necessary for input preparation, and the amount of time required for understanding the vast amount of output renders the use of inelastic time history analysis unfeasible for the assessment of seismic performance.

1.15 Design for Forced-Based Seismic Activity

In the current approach of Forced-Based code prescriptive seismic design, designers are bounded to select, proportion, analyze, design and detail the elements of building structure to satisfy prescriptive safety criteria of design code [1]. At first, element sizes along with the geometry of the structure are estimated from which member's elastic stiffness is estimated. Various assumptions are done in provide guidelines for the calculation of the suitable stiffness of the members. In some codes un-cracked stiffness is used, while in other codes reduced stiffness is used to reflect the effect of cracking.

Based on the stiffness of members, natural period of structure is determined using Equation (1).

$$T = 2\pi\sqrt{\frac{m}{k}} \quad \text{................ (1)}$$

Here, m & K are mass & stiffness of the structure, respectively.

In most of the building codes, structural period is calculated by Equation (2).

$$T = Ct\,(Hn)m \quad \text{............ (2)}$$

Where Ct is an unchanging factor that is selected based on structural system, Hn is the height of the building, and m is a constant that varies according on the types of materials and structural systems used.

The design base shear is calculated by Equation (3).

$$V = SaW \quad ....(3)$$

Here, S_a Does the fundamental seismic coefficient change depending on the seismic intensity and the soil conditions, building period, occupancy category, and ductility capacity. W is the building's seismic weight.

The structure's predicted base shear is spread throughout the structure to different levels (e.g. storey levels) of the building to analyze with the earthquake forces. Then, the moment capacities at required locations is determined and member sections are designed. Displacements, drift, or displacement ductility under seismic forces are estimated. If the displacements, drift, or displacement ductility are within the specified limit, then the design is finalized.

CHAPTER 2.
LITERATURE REVIEW

2.1 Overview

Earthquakes that happened all over the world, showed that seismic forces acting on the structures have damaged the structure causing loss of property and life. Present seismic design methods documented in the relevant design codes are incapable to predict the behavior of reinforced concrete structural components beneath periodical loading. The best alternative has emerged as performance-based seismic design (PBSD). to these code-placed procedures. In PBSD the building performance is expressed in terms of the structural and non-structural performance levels. The severity of the destruction is defined in terms of repairs, downtime, and casualties labeled as sudden occupancy, person welfare, and fall down elimination. Though these procedures can identify the state of damage, they are incapable to rectify the amount of destruction in a numeric value. In the past many researchers have put forth damage assessment methods, using damage indices which were formulated based on structural and response-based parameters. The present study attempts to brief all such efforts and explores the gray areas for future research. Besides, numerical ways to integrate the damage of the structure along with different assumption amount interpret in PBSD are explored.

The lesson learned from the past earthquakes has shifted the focus of the structural engineers towards attaining structural safety under seismic events. In India and its subcontinent, the present state of practice for the design of reinforced concrete buildings compelled to bear to lateral loads is force-based. In forced-based design, the inelastic behavior of structural components is taken care of by applying the limits of acceptable risk to material strength,

design loads, and serviceability. The force-based designed structures are capable to sustain repairable damages during minor and moderate earthquakes but were collapsed during severe earthquakes.

With the increase in the knowledge of earthquake engineering, predictive methods of design became common in practice [Ghobarah, 2001]. The most common practices followed under the predictive methods of design are; capacity-based, energy-based, stiffness-based, displacement-based approaches, and comprehensive design considering life cycle cost [Ghosh et al., 2011]. A framework covering all the above design approaches design for earthquakes that is centered on performance is called

(PBSD). The primary objective of the PBSD is to assess the structural performance in consideration of the design of seismic risks. [Yu-Chi Sung et al., 2006; FEMA 445, 2006].

In its current configuration, PBSD is primarily concerned with two different aspects: production-based seismic evaluation and performance-based seismic assessment. (The production of the building is assessed in PBSE by making a comparison between the capacity of the structure and the imposed request. On the basis of the amount of damage received by structural and non-structural components of the building for the intended displacement values, several performance levels of the structure are determined. These production levels are broken down into five categories: Operational Level (OP), Immediate Occupancy (IO), Life Safety (LS), and Collapse Prevention Range (CP) (C). This performance level provides a description of the method by which a structure may collapse when subjected to seismic stresses; nevertheless, they are unable to provide any damage value on a numeric scale. [ATC 40, 1996; FEMA 356, 2000; SEAOC 2000; FEMA 440, 2005; Boroujeni ARK, 2013; Zameeruddin and Sangle 2016; Zameeruddin and Sangle 2021a].

2.2 Performance-based Seismic Evaluation

The perspective design technique has been shown to be inferior to the performance-based seismic design (PBSD) approach, which has shown to be a superior alternative. There are several different PBSE approaches that have been offered through PBSD papers such as ATC 40, FEMA 356, ASCE 41, and FEMA 440. These publications have offered a variety of modeling and acceptance criteria for nonlinear modeling of reinforced concrete (RC) section geometries. [Zameeruddin and Sangle 2021a; Zameeruddin and Sangle, 2017b; Mondal and Ghosh 2013].

1. No destruction **2.** Slight destruction **3.** Average destruction **4.** Critical destruction or slump

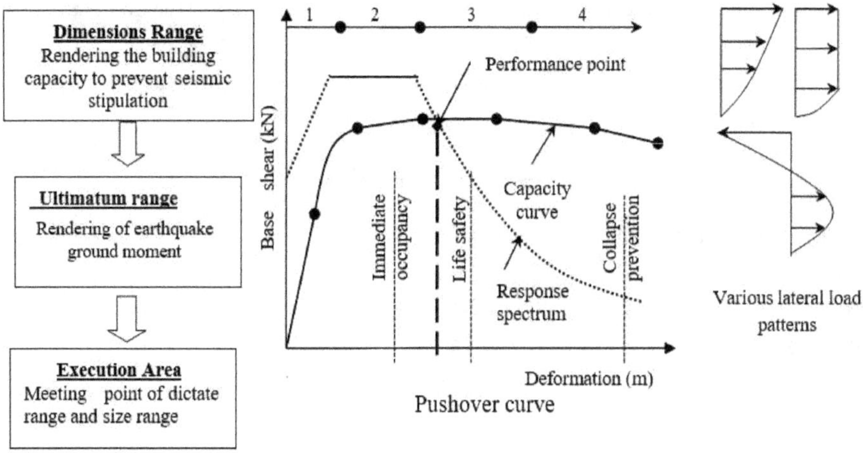

Flowchart of PBSD

Figure 2.1: ***Typical pushover investigation procedure***

Pushover analysis is used in the nonlinear static analysis processes offered in PBSD publications (POA). Using a predetermined pattern of lateral loads, POA gradually applies lateral stress to the structure until a plastic collapse mechanism has been achieved. Plastic hinges on the structural components are used to trace the propagation of inelasticity during POA as a lumped-plasticity technique. For evaluating a structure's nonlinear reaction, we utilize its capacity spectrum, demand spectrum, and performance point (shown in Figure 1) to determine its nonlinear response. [Zameeruddin and Sangle 2021; Zameeruddin and Sangle, 2017b].

The fundamental objective of PBSD is to conduct an analysis of the. seismic response of the structure within the pre-defined performance limits. These constraints are outlined in terms of performance. the acceptable risk of damage to structural and non-structural components. The engineering demand parameters (EDPs) used in defining these performance levels include; strain, stress, deformation, story displacement, damage (including crack width), and a force in a member [Zameeruddin and Sangle 2021a; Mondal and Ghosh 2013; FEMA 445, 2006]. Table 1 describes various performance objectives considered in PBSD.

Table 2.1: Damage states and drift restrictions are included in the performance levels.

Conditions	Damage state	Drift
(IO) Immediately occupied, fully operational.	Zero Damage	< 0.2%
Operational, Damage under control, Average	Repairable	< 0.5%
Life safety: damage condition, close to collapse, limited safety	Irreparable	< 1.5%
Hazard reduced	critical	< 2.5%
Collapse	-	> 2.5%

Methods such as the Capacity Spectrum Technique (CSM) and Displacement Coefficient Method (DCM) are part of the PBSE, which makes use of the nonlinear static method (NLSP) (DCM). In CSM, the capability of the structure (capacity spectrum) is compared to the imposed call.

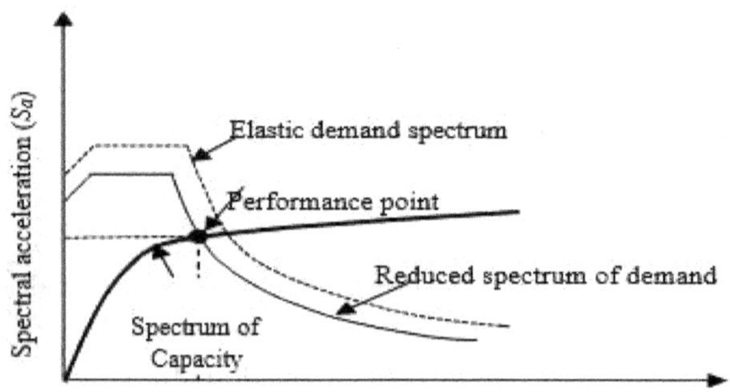

Figure 2.2 (a): ***Capacity spectrum method considering ATC 40***

on the building. The detailed meeting of above two arcs yields the output point, which estimated the buildings reaction. The CSM procedure is shown in Fig. 2(a). The Target dislocation in DCM is calculated using coefficient methods. The replacement of a featured node, specifically on a structure's rooftop, is referred to as target displacement. The DCM procedure is depicted in Fig. 2(b) [Zameeruddin and Sangle, 2016]. The CSM and DCM procedures had been improved by the following creation PBSD papers, FEMA 440 and ASCE 41. Zameeruddin and Sangle (2016) published a systematic study of these procedures in the literature.

2.3 Performance-based Seismic Assessment

In order to assess the functioning of the structure a damage index (DI) is needed to be defined [Kappos, 1997; Powell andAllahabadi, 1988; Brog and Rossetto, 2010]. The DI scales the damage state of a structure and its parts in a numeric value which ranges from 0 to 1. The value "0" shows the untouched condition and "1" describesthe collapse state. DIs are bifurcated into two classification Global DI and Local DI. Local DI shows the damage of individual members whereas global DI gives the weighted average of damages to all local DI at different structurallevels. Based on the EDPs used in the formulation of DI, DIs is further classified as response-based DI and structural parameter-based DI. These DI may be cumulative DI, Non-cumulative DI, and Combined DI [Zameeruddin and Sangle 2017a; Habibi and Izadpanah, 2013a; Habibi and Izadpanah, 2013b; Heo and Kunnath, 2013; Mihai Mihai*, 2013; Ghosh et al, 2011; Wang, J.-F. et al., 2007; Ghobarah, 2004; Ghobarah, 2000; Ghobarah et al, 1999; Powell, G. H. and Allahbadi, R., 1998; Kappos A. J., 1997; Williams, M. S. and Robert, R. G., 1995]. A comprehensive review of all DI's and their suitability are discussed in the literature by Zameeruddin and Sangle,2017a and 2021b.

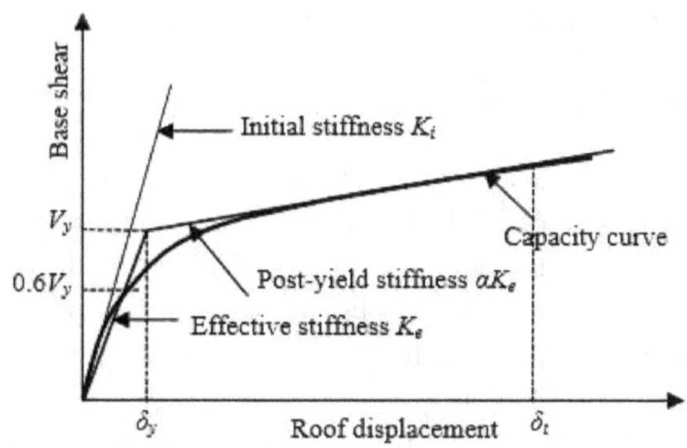

Figure 2.2 (b): ***Displacement coefficient method as per FEMA 273***

In RC structures, damage relates to inelastic deformations. That's why a DI includes deformation as a variable entity. The story displacement and inter-story displacement are used as EDPs for global DI. A tension is a member, angle at a cross-division; rotations at fragment termination are used to define local DI. Figure 3 illustrates the process of determination of the level of harm sustained by a member of the R.C. [Zameeruddin and Sangle, 2017a]. In past, There have been various attempts made by scholars to define Damage Indices (DIs) make use of arbitrary vigorous and unchanged survey results. Due to the complex analysis process, nonlinear dynamic processes were not common in practice. Comparatively nonlinear static procedures are common in practice, but addressing the cumulative effects still needed to be studied [Zameeruddin and Sangle 2021b].

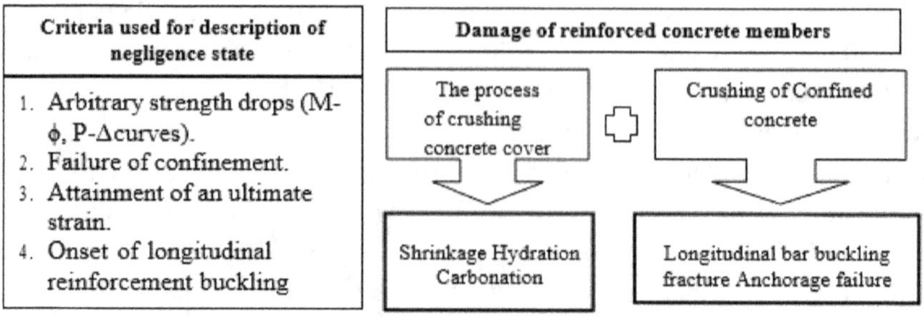

Figure 2.3: ***Identifying the extent of damage in RC members*** [31]

In its current configuration, PBSE procedures have potential to judge the plastic-yield mechanism of RC formation subjected to lateral loads, but they are incapable to provide a damage value. This study aims to put forth some DIs using the EDPs obtained as a result of PBSE performed on RC structures.2.4 Nonlinear Static Analysis.

2.4 Nonlinear Static Analysis

To achieve the desired displacement value, the RC formation put through to incremental edgewise forces in POA. The dormancy forces produced during a seismic event are represented by the lateral loads applied to the RC structures. Many researchers have provided various lateral distribution profiles. In a pushover study, the lateral force profile affects the structural response, so it's best to use a variety of edgewise force formation. [Ghorbah A., 2001] One of these ensembles can have upper and lower limits for edgewise forces. Gravity loads were used as distributed element loads in Stage I, which were calculated using the yield line theory. The secondary loads are also taken into consideration. For complete loads applied in a single stage, a gravity analysis was performed (i.e., force-control). In Stage II, the structure's state after Stage-I was finished was revisited. POA in Stage-II was used to apply lateral stresses in a sequential way.

In order to build a default plastic hinge's moment–rotation curve, FEMA 356's stress-strain relationship was used. Nonlinear frame parts, such as the beams and columns shown in Fig. 2.4, were modeled utilizing concentrated M3 and P-M3 plastic hinges on both ends. [Zameeruddin and Single, 2017b], the acceptance requirements for final spinning potential are labeled sudden tenancy (IO), life protection (LS), fall down interception (CP).

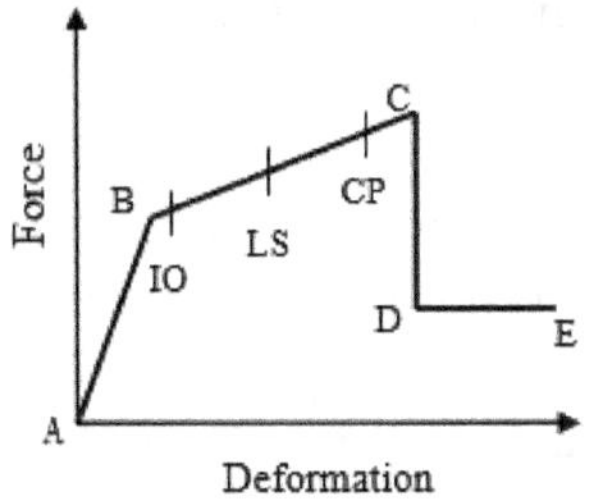

Acceptance Criteria

	M/M_y	Rotation/ SF		Plastic deformation /SF	
A	0	0	0	3E-03	0.01
B	1	0	0	3E-03	0.01
C	1.1	0.015	0.025	0.012	0.02
D	0.20	0.015	0.025	0.015	0.025
E	0.20	0.015	0.05	0.015	0.025

Figure 2.4: ***A typical plastic hinge's force–deformation relationship***

Ghobarah et al., (1999) elaborated that DIs may be divided into two basic groups: local DI and global DI. An individual member's damages are quantified by the local DI (e.g., at individual joints or an intended cross section). Weighted averages of local damage at various structural levels provide the global DI, which is a measure of the overall damage to the structure. Noncumulative, cumulative, and mixed DIs are all types of DIs. In earlier investigations, all of the known DIs and their limitations were thoroughly examined.

A. Ghobarah1, H. Abou-Elfath, and Ashraf Biddah (1999) showed an entirely new method for assessing structural damage has been developed that is more appropriate for non-linear structural analysis. Static pushover analysis may be used to evaluate the predicted damage to buildings in earthquakes of varying intensity, according to a realistic manner. The suggested approach for damage assessment provides simple, consistent, and reasonable damage indicators for structures, as shown by the results of the study of ductile and non-ductile reinforced concrete buildings.

Ahmed Ghobarah (2001) emphasized on the fact that the real dependability of the design in terms of accomplishing the goals is unknown at this time. The practice of seismic engineering in the future will be based on clear performance standards that can be measured, taking into consideration many levels of both performance and hazard. SEAOC Vision 2000, ATC 40, and FEMA273 and 274 are the three papers that are credited with establishing the groundwork for the principles of performance-based design. The publications made an effort to design methods that may be utilized as seismic requirements in building regulations and sought to do so. The advancements that are going to be made in performance-based design for seismic engineering will be directed toward a general design methodology that will allow performance-based design at multiple performance and hazard levels, and they will take into account the entire soil foundation structure system.

X.-K. Zou, C.-M. Chan (2005) proposes an efficient computer-based approach that automates the pushover drift performance design of reinforced concrete (RC) structures by including pushover analysis coupled with numerical optimization processes. This results in a more accurate and reliable design. It would seem that steel reinforcement is the most cost-effective material when compared to concrete materials. This material may be efficiently employed to limit drift beyond the occurrence of first yielding and to provide the needed ductility of RC building frames. The ratios of steel reinforcement are taken into consideration as design variables throughout the process of design optimization in this research. The nonlinear and inelastic seismic drift responses that are obtained by the pushover analysis may be clearly described in terms of the element design variables by using the idea of virtual work. In this research, an optimality criterion approach for addressing the explicit performance-based seismic design optimization issue for RC structures is described. The task at hand involves optimizing the seismic design. The efficacy and applicability of the suggested optimum design technique are shown using two building frame examples provided below.

Yu-Chi Sung, Chin-Kuo Su, Chuan-Wei Wu, I-Chau Tsai (2006) propose a new method with several adjustments is shown. It is anticipated that the evaluation would get more accurate while also being more efficient. The primary objective of the change is to consider the performance points on the structural capacity curve as input, and then to calculate the associated seismic demands of RC structures. This will be accomplished by modifying the structural capacity curve. After completing the seismic damage assessment processes that have been provided, a pushover analysis that is based on the given methodology is carried out in order to determine the maximum PGA that the building is able to withstand. After that, the NTH analysis is used so that the correctness of the findings may be examined and confirmed. It can be shown that the suggested technique produces findings that are satisfactory, and it is anticipated that this will be beneficial in both simplifying the process of assessing the seismic damage to RC structures and developing plans for repairing, upgrading, or reinforcing these buildings.

Mehmet Inel, Hayri Baytan Ozmen (2006) investigated the potential for a pushover analysis to provide different findings depending on whether or not the nonlinear component characteristics were predefined or user-defined. In the context of this research, low-rise and medium-rise buildings, respectively, are defined as structures with four and seven stories. In the user-defined hinge characteristics, it is assumed that the plastic hinge length and the transverse reinforcement spacing are effective parameters. According to the observations, the length of the plastic hinge and the spacing between the transverse reinforcements do not have any effect on the base shear capacity; nevertheless, these parameters do have a significant impact on the displacement capacity of the frames. According to comparisons, an increase in the quantity of transverse reinforcement results in an improvement in the displacement capacity. Although the capacity curve for the default-hinge models is fair for contemporary buildings that comply with codes, it is possible that it is not appropriate for other kinds of structures. Due to the fact that the majority of pre-existing structures in Turkey, as well as in some other countries, do not comply with the standards of contemporary code details, the installation of default hinges requires extra caution. The data make it abundantly evident that the user-defined hinge model is superior than the default-hinge model when it comes to accurately depicting nonlinear behavior that is consistent with the features of the element. However, if the default-hinge model is favored owing to its simplicity, the user should be aware of what is supplied in the program and should avoid misusing the default-hinge characteristics. The user should also be aware of what is provided in the program.

Erol Kalkan, Sashi, K. Kunnath (2007) In this study, we compare the NSP predictions of typical steel and reinforced concrete (RC) structures with benchmark answers from a complete set of NTH assessments to see how well they perform. In the analysis, the Adaptive Modal Combination process predicted peak response metrics including inter-story drift and component plastic rotations more consistently than the other NSPs studied.

K.A. Korkmaz and E. Irtem (2008) showed that by comparing the findings of time history analysis for reinforced concrete structures, a comparison of the new and prior two common performance-based analysis methodologies was made. In the study report, the Capacity Spectrum Method and Displacement Coefficient Method processes are thoroughly explained. A four-story reinforced concrete building is used as a test case to see how the various approaches compare. The chosen structure is located in an area with a high risk of earthquakes. Both the maximum inelastic displacement and building performance level may be calculated for each approach. According to the building's displacement requirements, the current and past methodologies' findings are summarized and compared with time history studies.

T.L. Karavasilisa, N. Bazeosb, D.E. Beskosb (2008) Presented Planar steel moment-resisting frames with vertical mass irregularity has their inelastic seismic response extensively parametrically studied. One hundred and thirty distinct earthquake ground movements were scaled to varying intensities in order to place the structures in a range of limit states in order to test the seismic and structural codes of Europe. So that an earthquake could be felt even when it didn't happen right next to a fault line, this was done. The number of storeys, the beam-to-column strength ratio, and the placement (top, midheight, and bottom) of the heavier mass all have an influence on the height-wise distribution and amplitude of inelastic deformation demands, according to the statistical analysis done on the response databank. There is no evidence that mass ratio influences the reaction. For a particular strength reduction (or behavior) component, three essential response values may be derived using nonlinear regression analysis, which reflects the aforementioned impacts. A building's maximum rotation ductility, maximum interstorey drift ratio, and maximum roof displacement may all be found here. For a specific strength reduction (or behavior) factor, these equations are utilized to construct simple formulae that represent the aforementioned impacts.

H. J. Jiang, L. Z. Chen, And Q. Chen (2011) indicated that an adjustment should be made to the original Park-Ang damage model in order to get rid of the non-convergence issue that it had at upper and lower bounds. The cyclic test results of flexure-dominant RC members from the database given by the Pacific Earthquake Engineering Research Centre and the author's own tests are used in the calculation of the combination coefficient for the updated model. According to the findings of the comparison between the modified model and the original model, the damage index of flexure-dominant RC members can be calculated by the modified model with more accuracy and less dispersion than the original model. The improved damage model computes the damage indices for each member while they are in primary damage states. Setting an individual limit value for the damage index is the method that is used to quantify the performance levels of RC structural parts.

Siddhartha Ghosh, Debarati Datta, Abhinav A. Katakdhond (2011) A performance-based seismic design must always include a comprehensive analysis of the possible structural damage that may occur. Because it is one of the seismic damage measures that provides the most accurate representation of structural damage, the Park–Ang damage index was chosen for use in this study as the seismic damage measure. Response spectra are the most prevalent kind of instrument that are used for the purpose of measuring the seismic danger that is present at a location. Response spectra also indicate the demand that is placed on single degree oscillators. Three similar single-degree system-based approximation approaches are provided in order to leverage these spectra for the estimation of the Park–Ang damage index demand on an MDOF system. These strategies are evaluated on three moment-resisting frames while subjected to a variety of ground motion conditions. When comparing the results of an analogous system scheme with the estimates obtained from a nonlinear response history analysis of the MDOF model, one may determine how successful the scheme is. The effectiveness of these strategies is evaluated based on global and storey-level damage indexes. In these case studies, variation of the non-dimensional parameter known as is also taken into consideration. In general, each of the three strategies seems to be successful, but to varied degrees depending on the situation. Because these systems may employ response spectra for demand estimate and minimize computing cost, they are suggested for use in damage-based seismic design and performance assessment of buildings. This is one of the reasons why the proposed approaches were developed.

A.R. Habibi, M. Izadpanah (2012) a novel design process that prevents damage to R.C. moment resisting frames has been created. In order to achieve this objective, first, a static damage criterion is created, and then, second, the new design approach is provided. The strategy that has been suggested is used in the construction of two frames made of reinforced concrete that are subjected to a total of seven earthquake acceleration recordings. In order to demonstrate that the approach being presented is reliable, an inelastic damage analysis will be carried out on the frames. It has been proved that the technique that has been suggested may be quite useful and practical when it comes to the design of RCMRFs with damage control.

Saurabh Shiradhonkar and Ravi Sinha (2012) For two sample buildings, the significant damage indices from the research literature are assessed by doing an analysis of the structures to see how earthquake loading affects them. In this work, several damage indices that are currently accessible are compared, and their applicability to the measurement of intermediate damage states in buildings is evaluated. It was discovered that the existing damage indices define states of the structure that are undamaged and collapsed, but they do not estimate intermediate damage states when the damage is mostly generated by beam damage and localized collapses.

Apurba Mondal, Siddhartha Ghosh, G.R. Reddy (2013) Estimating the actual values of this factor for realistic RC moment frame buildings that have been designed and detailed according to the Indian standards for seismic and RC designs as well as for ductile detailing is the primary focus of this line of research. These estimated actual values are then compared to the value that is suggested by the design code. According to the findings, the Indian standard prescribes a value of R that is far greater than the actual amount, which presents a safety risk. The behavior of the structure has not been verified by any nonlinear response- or time-history study. Although the many characteristics that were used in the work that was provided were thought to be deterministic, in actuality, the statistical fluctuations among them were big enough to necessitate the employment of a framework that was based on dependability for this investigation. The response reduction factor (R) value that is suggested in IS 1893 for 'ductile' and 'special' RC moment resistant frames has been subjected to an in-depth investigation for the purpose of determining whether or not it is accurate. The work that is being given here has taken into consideration four RC moment framed structures, each of which is situated in zone IV, has fundamental vibration periods that span a wide spectrum, and was planned and specified in accordance with the Indian standard guides IS 1893 and IS 13920. The emphasis has been placed on the following areas: a detailed modeling of the inelastic moment–curvature behavior; consideration of realistic design practices; a component-wise calculation of the factor R; consideration of realistic performance-based limit states at both the structure and member levels; and consideration of realistic design practices. According to Performance Limit 2 (limits on member rotation based on section dimensions and actual reinforcements), the suggestion provided by IS 1893 is on the more cautious end of the spectrum. This restriction does not, however, account for any structure-level behavior, such as interstorey drift, and this should be brought to your attention. In terms of R, the strong-column-weak-beam criteria that is used during design does not make a significant impact. The findings of the current research are restricted due to the fact that just a single plan configuration (without plan-asymmetry) in a single seismic zone was taken into consideration. This makes it impossible to draw any firm conclusions. In addition, the behavior of the structure has not been verified by any nonlinear response-time-history study. In the work that was provided, the many parameters that were employed were thought to be deterministic; but, in actuality, the statistical variances among these parameters are big enough to necessitate the employment of a framework that is based on reliability for this research.

Haoxiang He, Maolin Cong, and YongweiLv (2013) the adjusted coefficient, which is offered taking into consideration the cyclic load and hysteresis energy, was the one that was proposed. As additional forms of damage measurement, the number of inelastic cycles and the coefficient of hysteresis energy concentration are also included. As a result, several damage indices pertaining to displacement and energy are taken into consideration for performance-based design. The accuracy of the relevant data and assessment conclusions was confirmed by doing a nonlinear analysis on four distinct structures, which was then used to validate the fuzzy seismic damage assessment approach.

Heo and Kunnath, (2013) showed the fact that the damage models that are presently available employ nonlinear time-history analyses was addressed (NRHA).

The implementation of these models in damage assessment is, however, restricted due to the time-consuming and iterative process that is required in the calibration of these models. Very few academics have made an effort to define DIs by the use of nonlinear static techniques (NLSP).

ImaMuljatia, Fransiscus Asisia & Kevin Willyantoa (2015) assessed the performance of DDBD on a normal concrete special moment resistant frame and compared it to two types of FBD: the equivalent lateral force technique and the response spectrum analysis. Every approach was conceived by using the most recent version of the Indonesian seismic code, and it was validated by applying the identical method nonlinear time history analysis. Every design approach is executed in a single cycle of design without any attempt being made to increase the performance level in order to get a feel for how accurately each method can estimate seismic demand. Story drift, damage indices, and the structural failure mechanism are the metrics that are used in the process of assessing the performance of the structure. When it came to accurately anticipating tale drift, DDBD performed far better than FBD did. All approaches enjoy good damage indices level. Even though every method has a tendency to produce a poor mechanism, DDBD does not require any improvements because the structures already meet the performance goals that were set for them. On the other hand, FBD requires additional effort to improve the structural performance because the design procedure needs to be redone from the very beginning. In this regard, the DDBD technique is superior than the FBD in terms of its level of efficacy.

Mohd. Zameeruddin. Saleemuddin and Keshav K. Sangle (2016) investigated the stiffness-based damage index (DIc), which was subsequently presented as a straightforward formula and described as being based on the nonlinear response obtained from nonlinear static techniques. It is beneficial due to the fact that the pushover only has to be conducted once for the inertia loads derived from the corresponding static approach specified in BIS 1893. This helps to indicate the degree of damage that has been done to the structure in issue. Damage assessment of sample RC frames, each of which represents a distinct kind of construction, is performed with its assistance. The damage values are tied to FEMA356's drift-based damage index so that DIc may be extended to accommodate the various performance levels that are established by that standard. The findings indicate that DIc is consistent with drift damage values and is a useful instrument for use in real situations.

Mitesh Surana, Yogendra Singh, Dominik H. Lang (March 2018) When compared to the conforming structures, the non-conforming structures had a much lower collapse capability even though they were built with equal base shear coefficients. This illustrates the impact of a building's seismic collapse capability on the collapse mechanism. The collapse mechanism is heavily influenced by the seismic design SCWB ratio. Failure mechanisms only occur in buildings that don't meet the SCWB design criteria; on the other hand structures meeting the SCWB design criterion display beam failure mechanisms solely. Even in structures with a SCWB ratio of 1.4, column yielding couldn't be prevented. Interestingly. Beams in a single or multiple storeys failed in these buildings, and this led to the collapse of the structures. In the 8-storey buildings that were analyzed, this was the case. It seems that structures that do not meet the SCWB design requirement have unacceptably high risks of collapse (up to 60% in the case of mid-rise buildings) for the MCE hazard level. For buildings that meet the SCWB requirement, this performance dramatically improves. A seismic zone V 8-story structure constructed to Indian Standards that complies with the SCWB design criteria barely just makes it through the FEMA P695 standard for a 20 percent likelihood of collapse in the event of an individual building experiencing MCE hazard. In spite of this, the buildings that meet the SCWB standard function well as a group. A better capacity design approach might improve the collapse performance of RC frame structures, based on the reported modes of failure for the models under study. While further research on this issue may shed light on whether or not such enhancements are attainable without adopting changing SCWB ratios across the building's height, failure of beams at all levels is still a possibility. Structural models and plan forms are restricted in this investigation. The results can only be extrapolated to a restricted number of RC SMRF structures that are of a specified height. Furthermore, only far-field ground-motion recordings were employed in this investigation. The features of ground-motion data collected in the near field versus the distant field may vary dramatically. Because of this, independent investigations are suggested to investigate the influence of near-field earthquakes on the governing collapse mechanism and the collapse fragility.

Ajay Singh Thakur and Jagdish Chand Chandigarh University Garhuan (2019) The findings showed that the cross section of the beam and column matters a great deal in creating capacity-based design based on the strong column weak beam idea, which is systematically represented by the hinges generated during pushover analysis. The moment capacity ratio is critical in raising the column's ductility and moment resistivity in accordance with the code's recommendations. Hinge and beam mechanisms are used to build the frame; however, only the ground level column's bottom node is hinged. This means that shear walls are needed in order to reduce the building's displacement and enhance its structural integrity as the number of floors rises.

Rita Bento and Mário Lopes February (2020) Non-factored action-effects owing to the two specified actions were allocated to the end sections of beams and columns in this study's initial phase, as detailed in this publication. There were just two exceptions to this rule: The dead load and seismic load had quasi-symmetrical action-effects, which resulted in a tiny bending moment. Due to their extremely low ultimate curvature (proportional to the yield curvature), it would be unrealistic to use these sections to condition the collapse of an object. It was decided that the flexural capacity of all sections should be set at Mrd = 40kN.m, which corresponds to a small amount of flexible reinforcement, because in real-world situations the flexural capacity is never too low due to minimum reinforcement requirements. Action-effects are always smaller on top of the first-floor columns, leading to premature collapses of those sections below and above. Since real columns are never designed with weak points, this is untrue. This section's seismic bending moment was set to the same value as the bottom section's in order to remedy the situation. The development of plastic hinges in columns (sb1 and sb2) and beams (sb3 and sb4) was examined in both groups of structures (wb1 and wb2). To illustrate the extremes of actual design practice, these two examples were created. All of the non-plastifying elements had their moment capacities increased by a strength-over-strength factor of 1.35. Since the spring at the foundation is relatively stiff, the highest bending moment occurs at the base section of the wall. To ensure that the wall's elasticity continues above the base hinge, the same flexural capacity was assigned to all the cross sections' moment capacity. All structural members in this study had ductility that was greater than their actual tensile strength. Wall ductility may be less than that of linear members such as beams or columns with low axial forces, which is a real possibility. Consideration of this constraint, then, may lower the intensity factors even more when it comes to collapse-prone buildings, such as those built for the creation of hinges at beams.

Mohd. Zameeruddin, Keshav Sangale (2021) Using pushover, this section of the research aims to evaluate the seismic performance of MRFs exposed to various lateral stress patterns. The attention has been on the parameters such as basic time period, base shear, tale displacement, and component-wise calculation of response reduction/modification factor in relation to different performance levels. According to the findings of this research, lateral load patterns employed to depict inertia loads generated by earthquakes are necessary for the Pushover analysis. It is recommended that inertia loads be measured using a set of lateral load patterns to establish upper and lower limits. IS 1893 lateral load pattern Push 1 specifies lower limit values, uniform lateral load pattern Push 2 defines upper bound values, and elastic first mode lateral load pattern Push 3 defines middle values, according to analytical investigations. Story displacement and interstory drift were analyzed to see whether Push 2 load case underestimated story and interstorey displacement compared to Push 1 and Push 3 load cases. The regulation mandates that interstorey distribution should be consistent throughout the building's height.

Mohd. Zameeruddin, Keshav Sangale (2017) The example MRFs in this work were subjected to SAP 2000 V17 displacement-controlled NLSP. Each MRF has a target displacement of 4% of the frame height [ATC 40, 1996]. There were two steps to the analysis: one for gravity loads and the other for the most prevalent lateral loads. Using the yield line theory and focused loads from secondary beams, gravity loads were used as distributed element loads in Stage I. In a single step, the complete gravity load was analyzed (i.e., force-control). In Stage II, the structure's state was retrieved from this analysis. Step-by-step nonlinear static analysis was used to apply lateral stresses in Stage II. Pushover analysis relies heavily on the lateral force profile; hence a series of lateral load patterns were utilized. A uniform load pattern, as stated in the codes or 7 elastic mode-1 pattern, made up one set, and inertia load upper and lower bounds were supposed to be derived from this. The second set had a primarily lateral load pattern. The IS 1893 lateral load pattern and the elastic first-mode lateral load pattern were the two most often employed lateral load patterns in this investigation. Figure 3 depicts the applied load pattern as a function of time. Each load combination in this investigation was examined for consequences of P - geometric nonlinearity. It was found that nonlinear frame parts like beams and columns may be approximated using concentrated M3 plastic hinges and P-M3 plastic hinges on both ends. In accordance with FEMA 356 requirements, parameters and acceptance criteria for modeling were employed. The IO, LS, and CP acceptance requirements for ultimate rotation capacity are shown. Modeling parameters and numerical acceptance criteria are outlined in this section. Cross-sectional area, rebar % in tension and compression, design shear strength, and axial loads are all factors in these characteristics.

M.S. AJ-Haddad and G.H. Siddiqi (2018) "A Study Leading to Preliminary Seismic Design Criteria for the Kingdom," a research study just finished, was dubbed "A Study." These design guidelines were developed as a result of a rigorous analysis carried out for this project. The iso-acceleration map generated throughout the course of the research was used to delineate seismic zones, and the most current state-of-the-art was used to estimate the zone factor for different zones. In most cases, the suggested design criteria are established in accordance with SEAOC and UBC regulations, following careful evaluation and adjustment of the relevant aspects. The map of the seismic zone and the assessed values of several seismic design parameters relevant to the Kingdom of Saudi Arabia are presented in this study. Seismic design of new and existing buildings relies heavily on this knowledge, as well as the simplified direction offered in this article to relevant sources.

Jiro Takagi, Akira Wada (2018) emphasizes the possibility that a vast number of structures, including whole cities, may be severely damaged or destroyed as a result of big earthquakes. There have been previous major earthquakes in which well-designed and built buildings that didn't collapse were destroyed rather than restored. Because of this, the earthquake-resistant design philosophy created in the preceding century should be updated to satisfy current social and economic needs and the Sustainable Development Goals (SDGs)... For contemporary and resilient civilizations, the seismic design philosophy for buildings and infrastructure should shift from life-saving to business continuance. Designing structures such that they can be swiftly and cheaply rebuilt after a severe earthquake is a key component of earthquake resilience.

Akshay Ahirwal, Kirti Gupta, Vaibhav Singh (2019), Multi-story buildings' seismic response was researched in detail, and it was found that diaphragm apertures significantly reduced the base shear, which in turn attracted lower seismic forces. The right placement of apertures may provide the structure with the necessary strength and serviceability. For both new and existing buildings, it's important to know how well the structures can survive disasters. Openings in diaphragms, which are typical for stair cases, architectural lighting, and more, may produce stress at discontinuous junctions with other building components, such as walls, floors, and ceilings. Discontinuous diaphragms are made without stress calculations and are deemed sufficient regardless of the impact of any gaps. Seismic response differences between two buildings with and without diaphragm discontinuity have been studied in this article. The seismic response of existing buildings with diaphragm discontinuity will be compared to that of buildings without it using CSI SAP200 software in the paper. Engineers have completed a thorough linear static and dynamic study. The modal time period, base shear, storey drift, and joint displacement results of the two buildings were compared.

Mathieu Giloulbe, Fouad Adnan Noman Abdullah Al-Shaibani, Abass Saad Lina (2019) related to PBSD in concrete structures and the implementation of this strategy. For structural and non-structural components, PBSD can provide more precise information about their performance levels. Methods. On a reinforced concrete irregular frame, this research applies Performance-Based Seismic Design. Pushover analysis was performed in order to carry out this task. Using story drift ratios as deformation limits, the performance thresholds for different seismic hazard levels may be defined. Seismic load-carrying capacity is increased by using Performance-Based Seismic Design, which achieves both performance and cost-effectiveness in one fell swoop. For different levels of earthquakes, it is also feasible to infer that PBSD derived by the aforementioned approach meets the acceptance requirements for immediate occupancy and life safety limits.

Mohd. Zameeruddin, Keshav K. Sangle (2019) moment-resisting frames are assessed for seismic damage. The nonlinear responses at the output of nonlinear static processes are used to create some new damage indicators. Newly specified static indicators coexist here with potential expansions of previously established static indicators. Formulating the damage involves the use of four different types of damage variables: displacement, acceleration, and shear at the base. Plastic ductility, loss of strength and stiffness deterioration as well as energy dissipation are addressed by the new damage indicators. In order to determine different degrees of performance, distinct drift criteria's damage value fluctuations were taken into account. The author sought to solve the constraints of the pushover analysis in this study by incorporating damage indicators that take cumulative effects into account while assessing plastic ductility, strength degradation, stiffness degradation, and dissipated energy of the structure. These new DIs make it possible to assess damage at any point along the capacity curve, regardless of displacement or force.

Archana J. Satheesh B. R. Jayalekshmi Katta Venkataramana (2019) Buildings with variable aspect ratios, different mass locations throughout the height, and also a specified level of in-plan eccentricity were evaluated for their seismic reaction. Earthquake loads are applied to building frames with up to 15 floors of vertical and plan eccentric mass integration. The three-dimensional finite element models of building frames are subjected to El-Centro ground motion in order to perform transient analysis, and the torsional behavior is assessed in terms of change in base shear, fundamental natural period, roof deflection, and ground rotation. When imperfections are present in the bottom half of the frames, the in-plan eccentricity of the mass irregularities has the least impact on the seismic response. The research presents a metric to measure the irregularity of the building's mass in relation to its location and in-plan eccentricity.

Chien-Kuo Chiu, Hsin-Fang Sung, Kai-Ning Chi and Fu-Pei Hsiao (2019) experiments on six RC column specimens with flexural, shear, and flexural shear failure modes were utilized to determine residual seismic capacity of damaged column members for specified damage states in this paper. Experiment data and associated studies are also analyzed to determine the decrease factors of strength, stiffness, and energy dissipation capacity for damaged RC column members, respectively. Seismic reduction factors for RC columns are proposed here based on damage statuses. Reduction factors for strength (0.5), stiffness (0.6), and energy dissipation capacity

(0.1) are given for an example of an RCC column with flexural– shear failure. The research presented in this study also presents a technique for assessing the residual seismic performance of earthquake-damaged reinforced concrete buildings (RCBs). To show the post-earthquake evaluation of the seismic performance of a damaged RC structure, this study uses one genuine earthquake-damaged school building.

S. Rana, M. A. R. Bhuiyan (2020) done the designing of reinforced concrete (RC) building frames following performance-based earthquake engineering approach considering soil-structure interaction (SSI) effect. SSI plays a significant role as it may have beneficial or detrimental effects on the seismic response of the superstructure. An archetype RC building has been analyzed and designed following the seismic design approach of the updated BNBC (BNBC 2020). Nonlinear time history analyses have been performed using suitable ground motions corresponding to selected performance objectives. Story drift, an indicator of damage, has been selected as a response parameter to quantify performance. The effects of base flexibility on the responses of the building in force-based and performance-based design approaches have also been assessed. The method suggested in this paper will help designers, owners, and stakeholders to make intelligent decisions in designing new or strengthening existing buildings to achieve the required performance of the structures. This paper has illustrated the gap between the typical code prescriptive forced-based and the performance-based design for seismic hazard. A typical reinforced concrete framed building has been designed with the code-based method and then assessed for various performance levels following the performance-based approach considering the soil-structure-interaction effect. The results from the NTHA with the SSI effect show that, the SDR value is increased only for Northridge ground motion records both for DBE and MCE hazard level. From the performance-based seismic assessment of the code-designed building it has been found that building is not adequate enough to achieve required seismic performance level corresponding to defined earthquake hazard level. However, the building should be analyzed with more ground motions to understand the response of the building under dynamic loading. PBSD approach emphasizes on performance of the structures with sufficient better confidence of the designer, owner, user and decision-makers to make the structures seismically resilient. Thus, it will be better to follow the reliable performance-based approach to design the building for seismic forces.

Mohd. Zameeruddin, Keshav K. Sangle (2020) This performance-based seismic design focuses on quantifying and characterizing the related structural damage in order to include them directly into the design or performance assessment approach, and both of these issues were solved. To address these problems, this work uses nonlinear static techniques to evaluate the performance of reinforced concrete frames. Seismic rules in India dictate that buildings must have 15-moment resistant frames, which were tested against diverse load patterns. It was determined that the seismic performance met several performance standards by comparing it to fundamental periods, roof displacements, interstoried drift ratios, base shears, and modification factors. According to Indian seismic code rules, the obtained findings were in conflict with the basic time period, upper and lower bounds of base shear drift ratio and modification factor.

2.5 Conclusion based on Literature Review

Post-earthquake behavior of RC frame demonstrates the inadequacy of the ERD procedure: described in present seismic code to account for inelastic behavior and cycle loading effects PBSD has established itself as the most attentive approach and effective answer to the issue. In PBSD, the desired behavior of the building is evaluated according to set of predefined performance objectives, which were formulated at the conceptual stage and followed by numerical investigation of various seismic demands with the development of advanced nonlinear analysis techniques, PBSD is becoming more commons; however the following problems must be solved:

1. The current state of practice describes performance objectives regarding discreetly defined performance levels, namely immediate occupancy, life safety and prevention based on transient and permeant drift, which does not clarify the associated repair cost, life loss, down time. Thus, these losses must be communicated in a more meaningful and useful manner to owners, designers and building officials.

2. The present condition of affairs in the field provides performance assessment approaches, which are simple in terms of their approach from elastic to inelastic methods of analysis, but the accuracy in predicting the actual building response and conservation in acceptance criteria is questionable.

3. The current state of practice provides performance assessment procedures for predefined performance levels without considering indicators Thus framework that properly accounts for adequately integrates all available damages models proposed by many scientists in PBSD Procedure must be developed.

Through an analysis of MRF's performance, this research makes an effort to answer the key problems raised, designed for gravity loads. The strategy is to prepare non-linear model of this structure with the installation of plastic hinges on the beams and columns.

In the past, many efforts were taken to scale the damage sustain by the structure or its components using Engineering Demand Parameters. This EDPs may be tension, a strain, a displacement, or stiffness value measured before and after subjecting the structure to seismic hazards. These EDPs were used to evaluate structural parameters and response-based parameters, scaled numerically from zero to one. The process is termed as PBSA. Many researchers have taken efforts to integrate PBSE and PBSA procedures to have a collapse mechanism and yielded damage value for a complete understanding of how the structure performs in design seismic event [Zameeruddin and Single, 2020b]. The present study aims to obtain state-of-art-of- review of the developments in the PBSD and PBSA. Also, an attempt has been made to put forth the possible integration of PBSE and PBSA, to state the probable performance and damaged state of the structure when subjected to a design seismic event.

CHAPTER 3.
RESEARCH METHODOLOGY

3.1 Insights

Recent earthquakes that have struck a number of nations all over the globe have brought to light the urgent need for a significant modification to the existing seismic design technique. The current seismic design guidelines are force-based, which means that forces and displacements are computed while staying within elastic limitations, and their combination is utilized to design the structural and non-structural components of a building. Checks for serviceability are carried out with the assistance of displacement limits and ductile detailing. The calculation of inelastic responses involves using the response reduction factor, which is related to the amplification of force or displacement. However, since this method is indirect, it leads to inaccurate estimations of the real building reaction.

The design approach is centered on the predictive way of analyzing possible seismic performance. This method is known as performance-based seismic design, and it is utilized with the intention of communicating the safety-related judgments (PBSD).

Design criteria are described in terms of attaining stated performance goals when the structure is exposed to the specified levels of seismic danger. PBSD is a broad design philosophy A realistic and accurate assessment of the danger to life, occupancy, as well as economic damage that may be caused by future seismic occurrences is required for buildings to be designed and constructed within the terms of the PBSD. There are several stages in the development of a PBSD project, all of which are iterative until the desired level of performance is achieved, the selection of performance objectives and continuing through preliminary design (based on those stated performance objectives), assessment (to determine whether the design meets those performance objectives), and redesign (if necessary).

In contrast to the force-based design (FBD), the performance-based safety design (PBSD) offers a methodical approach to evaluating the capacity of performance that may be found in a structure, a system, or its individual components. The technique offers a framework that may be used to determine the degrees of safety and property protection needed for the project, as well as the costs that are acceptable to the owner, the designer, and the building authorities in accordance with the unique requirements of the project.

The advantages of PBSD are as follows:

1. A more self-assured approach to architectural design.

2. Creating individual structures that are more cost-effective to create while still achieving the specified performance goals.

3. Designing particular structures to perform better than expected based on the current seismic standards is a third option.

4. Buildings beyond the seismic code-prescribed boundaries in terms of layout, materials, and systems, must be designed to satisfy the intended performance of current seismic codes.

5. A seismic hazard might cause significant damage to existing buildings, therefore it's important to figure out how much it would cost to repair such buildings.

6. Code-based seismic design criteria may be improved by assessing the current seismic code requirement for new buildings and determining its potential performance.

PBSD describes performance levels as a function of displacements and drifts rather than absolute values. It is possible to link a structure's damage condition with the degree of performance. This concept has resulted in a new design strategy based on the performance level of a building. The permissible amount of damage to structural and non-structural components is described as a building performance level.

Federal Emergency Mitigation Agency (FEMA) and Applied Technical Council (ATC)has put forth various building performance levels in their publications namely, FEMA 273, FEMA 356, FEMA 445, ATC 40, and ATC 58. Table 3.1 shows brief description of these performance levels.

Table 3.1: Building Performance Levels 273/356

Non-structural Performance Levels	Structural Performance Levels					
	SP-1 (IO)	SP-2 (DCR)	SP-3 (LS)	SP-4 (LSR)	SP-5 (SS)	SP-6 (NC)
NP-A (Operational)	1-A	2-A	NR	NR	NR	NR
NP-B (Immediate Occupancy)	1-B	2-B	3-B	NR	NR	NR
NP-C (Life Safety)	1-C	2-C	3-C	4-C	5-C	6-C
NP-D (Hazard Reduced)	NR	2-D	3-D	4-D	5-D	6-D
NP-E (Not Considered)	NR	NR	3-E	4-E	5-E	NR

With the increase in knowledge related to seismic hazards and structural performances under seismic loading, people across the world are concern about their structure performance in relation to direct and indirect losses. Direct loss includes damages(repairs) and life safety, whereas indirect losses relates to downtime (unused life span) of a structure.

At present in India, the practicing engineers are focusing on static (indirect) approach to address seismic forces. These procedures are not enough to justify the structure performance.What we need is to modify these procedures and proposed new method/ process of design highlighting key performance parameter. Beside this the proposed procedure should be user friendly to be practice in common. The finite element software such as ETABS, SAP, ABACUS, IDARC, DRAIN AND STAAD made it possible to analyse and observe nonlinearprocedure. They have been limited for some identified projects or research findings. With thisproject we aim to provide a procedure which will be based on this software and easy to practice by common individual. Beside this guideline process will help an individual to take appropriate measure against a predict failure.

3.2 Objectives of Study

Performance-based seismic evaluation (PBSE) methods and damage assessment ideas are the focus of this investigation. In PBSD documents, nonlinear static analysis underpins PBSE operations. With the development of advanced nonlinear analysis techniques computing tools, PBSE has become common in practice, but has a limited approach towards damage quantification. To appear as a well-defined PBSE, there is a need tohave a clearance on some issues as;

1.The performance is characterized by the present stage of practice. objectives regarding discretely defined performance levels, namely immediate occupancy, life safety, and collapse prevention based on transient and permanent drift, which themselves are incapable to clarify the associated repair cost, life loss, and downtime. Thus, these losses must be communicated in a manner more meaningful and useful to owners, designers, and building officials.

2.The current state of practice provides performance evaluation techniques, which are simple in terms of their approach from elastic to inelastic methods of analysis, but the accuracy in predicting the actual building response and conservatism in acceptance criteria is questionable.

The performance is characterized by the present stage of practice procedures for predefined performance levels without considering damage indicators. Thus, a framework that properly accounts for and adequately integrates all available damage models proposed by many scientists in the PBSD procedure must be developed.

3.3 Problem Statement

The performance-based seismic design has two key goals: properly defining the related structural damage for direct inclusion into design or performance assessment methods, and correctly estimating the uncertainties involved with the performance evaluation process. To determine a building's performance levels such as immediate occupancy, life safety, and collapse prevention, the performance evaluation procedures that are described in first generation (SEAOC vision, ATC 40, FEMA 356) and second generation (ASCE 41) performance-based seismic design documents use an inter-storey drift ratio and plastic rotations. These performance evaluation procedures were developed by SEAOC, ATC, and FEMA. It is necessary to associate a damage state with performance levels because these measurements, while they do provide information on the deformation of the elements and the displacement profile at critical states, are not adequate on their own to provide an assessment of the state of damage or the proximity of collapse.

A damage state is expressed with the help of damage indicator involving various engineering demand parameters. These engineering demand parameters may be generated via nonlinear analytical approach, the more rational approach is nonlinear static procedure (pushover).

Plastic ductility, loss of strength, stiffness deterioration, and energy dissipation are defined using nonlinear responses from the result of pushover analysis in this study. The drift-based damage index was used to expand the introduced damage index to different performance levels indicated in performance-based seismic design guidelines.

3.4 Overview

PBSD, which takes into account inelastic behavior and cyclic loading effects, has emerged as the best solution to the shortcomings of force-based seismic design approaches detailed in contemporary seismic models. Buildings may be designed with a realistic and accurate knowledge of the danger, to life, occupancy, and economic loss, that may arise from future earthquakes. Both the performance assessment process's uncertainties and the structural damage it causes may be quantified using PBSD and then used directly in the design or performance evaluation technique (Ghobarah, 2001; Zameeruddin and Sangle, 2016).

Objective of this research is to analyze the PBSE processes for reinforced concrete frame provided in several PBSD publications the reinforced concrete frame's maximum inelastic displacement and performance level are calculated using PBSE techniques and time history. While prior research studies have focused on the pros and disadvantages of performance assessment methodologies, the current study used gravity-based design MRFs and a performance index to investigate if various performance levels were associated with user-defined hinge and default hinge placements.

3.5 Available Performance-based Seismic Evaluation (PBSE) Methods

Nonlinear analytic approaches have made it feasible to get more accurate results from a complicated seismic study. The most often utilized PBSE methods for seismic analysis including computation in PBSD are the first generation (ATC 40 and FEMA 273), second generation (FEMA 356), and third generation (ASCE 41 and FEMA 440) PBSE procedures.

These two approaches, "Capacity Spectrum Method" and "Displacement Coefficient Method," are often used in nonlinear static seismic analysis. In these processes, the lateral load pattern is monotonically increased until the desired displacement is obtained (i.e. Pushover analysis). Weaknesses and failure mechanisms in the structure are revealed as the load increases. In order to account for load reversal and cycle behavior, a modified monotonic force-deformation criterion is used with damping approximations (Sharma et al., 2005).

Using pushover analysis, depiction of the displacement demand on the structure owing to the stated degree of seismic danger, lateral force resistance capacity curve and a comparison of imposed demand and structure's capacity, as illustrated in figure 3.1, is used to verify acceptable performance. These processes aid in determining which component of the structure is most likely to fail. Figure 3.2 depicts the emergence of a collapse mechanism as a consequence of the components (beam, column, etc.) beginning to yield and deform inelastically under increasing load and movement. Building capacity is shown in an easy-to-understand graphic curve.

Each PBSE method has provided a procedure to define the target displacement, and locate the performance point to identify the inelastic behavior. In this research, the main goal is to find out and to evaluate the available PBSE methods documented in PBSD and using their results estimation of the performance index, response reduction factor, interstorey drift, and A wide range of hinge arrangements and structural stiffness variations.

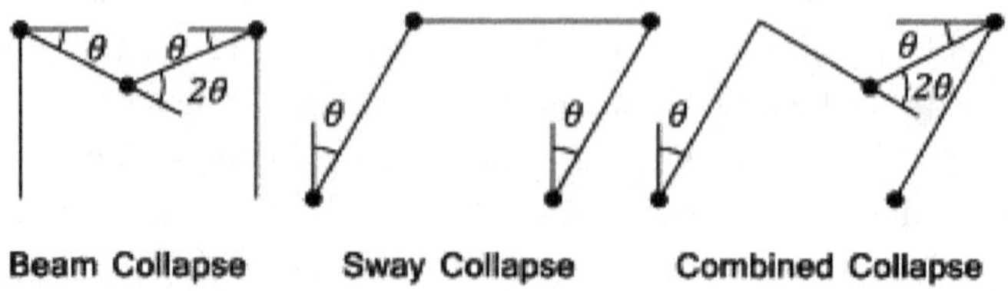

Figure 3.1. ***Formation of collapse mechanism at yield***

3.6 Example Moment Resisting Frame

The response and damage states of MRFs representing G+5 framed building was examined. Nonlinear static analysis was performed on the sample MRF (POA). First, second, and third generation procedures were utilized to assess example MRF's response. Analytical modelling of the sample structures has been carried out with SAP 2000 V 21.0. The example MRF was designed using IS 456:2000, IS 1893:2002 (Part 1), and IS 13920:1996. There were lateral stresses on the frame. The MRF may be found in seismic zone V, which has a z-value of 0.36, and it has a relevance factor of 1. (soil type). Structural elements are designed using a variety of materials, Figure 3.2 depicts the top views of an example MRF, as stated in Table 3.1. Table 1 displays information on the Material Characteristic of RC sections.

The illustrative MRF illustrates a medium-rise G+5 frame structure. This MRF was constructed in accordance with IS 456: 2000 (Rev), IS 1893: 2002 (Part 1), and IS 13920: 1996. Figure 3.2 depicts the usual configuration of the illustrative MRF. This MRF shows a commercial structure situated in seismic zone IV, according to IS 1892, on a medium soil type. It is assumed that the structural important factor (I) is 1.5 and the response modification factor (R) is 5. The height of the bottom level is 4.1 meters, while additional floor heights are 5 meters, and the span of the beam is 7.5 meters. The distance between each frame is 7.5 meters.

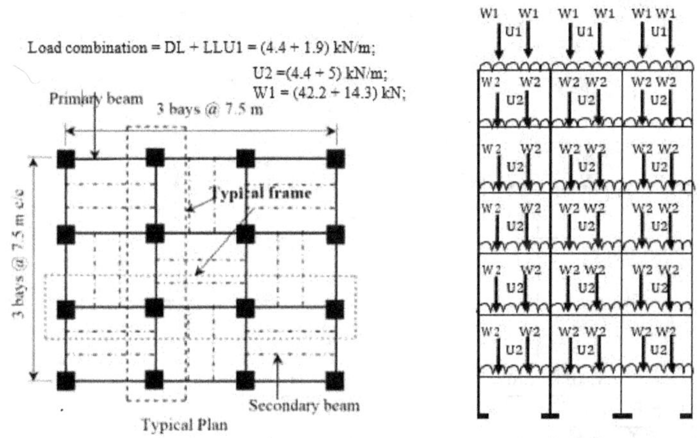

Typical elevation
(Symmetrical loading along X & Y direction)

Figure 3.2: ***Plan and elevation of example MRF (3B6S)***

According to IS 875 (Parts 1 and 2, 1987) and IS 1875 (2002), dead loads, live (imposed) loads, and seismic loads were included for the study of chosen MRF. Table 3.1 displays the seismic load estimate for the structure. The example MRF is exposed to a typical dead load (including finishes) (as seen in Fig. 3.3) and a live load intensity of 4 kN/m^2 for typical floor and 1.5 kN/m^2 for roof.

Table 3.2: ***Total seismic load on structure***

Sr. No	**Description**	**No.**	**Length (m)**	**Breadth (m)**	**Height (m)**	**Load intensity /Density kN/m^2 Or kN/m^3**	**Weight (kN)**	**Sub-total Weight(kN)**
A	**Roof (Terrace)**							
1	Dead weight of slab	1	22.5	22.5	0.1	25	1265.63	
2	Finishing load on slab	1	22.5	22.5		1	506.25	
3	Water proofing	1	22.5	22.5		2	1012.5	
4	Super imposed load	1	22.5	22.5		0	0	
5	Dead load of beams							**5593.08**
	Main Beam	8	22.5	0.3	0.6	25	810	
	Secondary Beam	18	7.5	0.2	0.5	25	337.5	

6	Columns							
	C1	16	0.5	0.5	2.50	25	250	
7	Walls:							
	Parapet Wall	4	22.5	0.245	1.00	20	441.00	
	Wall	4	22.5	0.245	2.20	20	970.20	
B	**Storey No. 6-5-4-3**							
1	Dead weight of slab	1	22.5	22.5	0.1	25	1265.63	
2	Finish load of slab	1	22.5	22.5		1	506.25	
3	Super imposed load	1	22.5	22.5		2	1012.5	
4	Dead load of beams							
	Main Beam	8	22.5	0.3	0.6	25	810	
	Secondary Beam	18	7.5	0.2	0.5	25	337.5	
5	Columns							**6372.28**
	C1	16	0.5	0.5	5.00	25	500	
6	Walls:							
	W1 outer	4	22.5	0.245	4.40	20	1940.40	

Sr. No	**Description**	**No.**	**Length (m)**	**Breadth (m)**	**Height (m)**	**Load intensity /Density kN/m^2 Or kN/m^3**	**Weight (kN)**	**Sub-total Weight (kN)**
C	**Storey No. 2**							
1	Dead weight of slab	1	22.5	22.5	0.1	25	1265.63	
2	Finish load of slab	1	22.5	22.5		1	506.25	
3	Super imposed load	1	22.5	22.5		2	1012.5	
4	Dead load of beams							**6128.83**
	Main Beam	8	22.5	0.3	0.6	25	810	
	Secondary Beam	18	7.5	0.2	0.5	25	337.5	
5	Columns							
	C1	16	0.5	0.5	2.50	25	250	
	C2	16	0.5	0.5	2.05	25	205	
6	Walls:							
	Wall 1	4	22.5	0.245	2.20	20	970.20	

	Wall 2	4	22.5	0.245	1.75	20	771.75	
D				**Plinth**				
1	Dead load of beams							
2	Main Beam	8	22.5	0.3	0.6	25	810	
3	Columns							
	C1	16	0.5	0.5	2.05	25	205	**2020.30**
	C2	16	0.6	0.60	0.55	25	79.2	
4	Walls:							
	Wall 1	4	22.5	0.245	1.75	20	771.75	
	Wall 2	4	22.5	0.245	0.35	20	154.35	

IS 456 (2000) criteria were used to develop the RC design for these MRFs. The RC section's ductile (seismic) detailing was done in accordance with IS 13920. Table 3.2 lists the different material attributes utilized in structural member design. Selection of RC member sizes was done in accordance with well accepted industry standards. The cross section of every column and beam in a certain storey is the same. Sections of reinforced concrete are shown in detail in Tables 3.3 and 3.4.

Table 3.3: ***Material properties of the design of example MRFs***

(IS 456, 2000; IS1786, 2008)

Material property	**Concrete M 25 grade**	**Steel Fe 415 grade**
Weight as a ratio of volume to weight (kN/m^3)	25	76.97
Weight per cubic meter (kg/m^3)	2.548	7.849
The elasticity modulus (kN/m^2)	25E+06	2E+08
Characteristic strength (kN/m^2)	25,000 (for 28 days)	415,000 (yield)
Minimum tensile strength (kN/m^2)	–	485,800
Strength of the expected yield (kN/m^2)	–	456,500
Required tensile strength (kN/m^2)	–	533,500

Table 3.4: ***A synopsis of the column section that was created for the model MRF***

Stories	Dimension(mm)	Exterior span rebar's (mm^2)	Interior spanrebar's (mm^2)
1	500 x 500	3214	2514
2		3073	2207
3		2899	2108
4		2627	2206
5		2195	2204
6		2000	2691

Table 3.5: ***A synopsis of the beam section that was created for the model MRF***

Stories	Dimension(mm)	Exterior span rebar's (mm^2)			Interior span rebar's (mm^2)		
		Top	Bottom	Top	Top	Bottom	Top
1	300 x 600	2351	1375	2544	2406	1292	2406
2		2620	1364	2690	2587	1298	2587
3		2568	1363	2595	2535	1301	2535
4		2388	1366	2394	2374	1301	2374
5		2102	1346	2086	2116	1310	2116
6		1070	865	1317	1309	731	1309

3.7 Nonlinear Modeling of Reinforced Concrete Members

The example's nonlinear reaction may be estimated. Analysis of the frame's nonlinear behavior has a substantial impact on the MRF. Bending, shear, axial tension, compression, and other failure modes for reinforced concrete members exist. Moment-curvature properties, axial force–moment interaction and shear force deformation characteristics are used to model the full behavior of RC members. Members' moment-rotation behavior determines the frame's nonlinearity; this relies on the plastic hinge section's moment-curvature properties and hinge length. In terms of plastic rotation capacity, these two factors also establish the "component level" performance limitations (Mondal, 2013). The stress-strain relation for concrete and steel may be used to calculate the theoretical moment-curvature analysis of the RC section. Axial force-moment hinges (P-M) and rotation hinges *(M- ϕ)* are also studied in this research (Sunitha et al., 2016; Ravi kumar et al., 2014; Sharma et al, 2005).

3.7.1 Rotation hinges (M-ϕ):

The moment curvature relationship is a function of section details. In order to generate this property, the reinforced concrete section is divided into layers to which the inelastic behaviorof either steel or concrete is considered. A trilinear elasto-plastic behavior of concrete and steel is considered figures 3.3 and 3.4 illustrate this point.

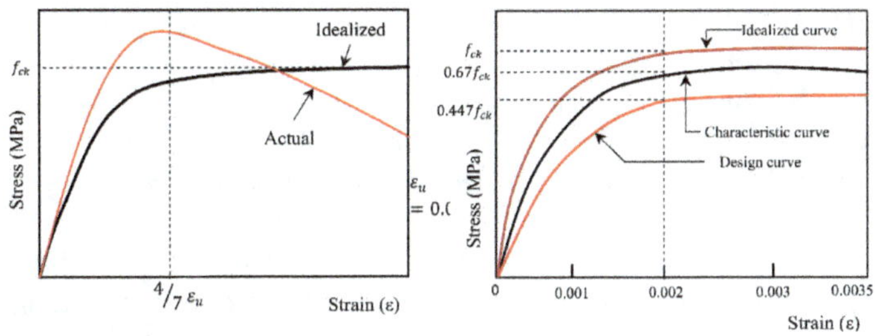

Figure 3.3: ***Stress-Strain curves for concrete (IS 456: 2000)***

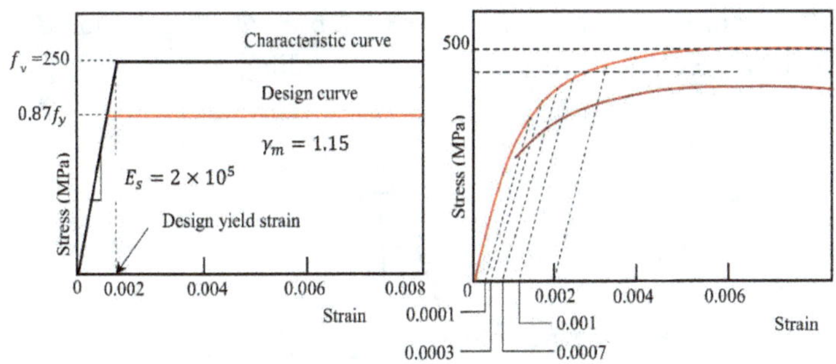

Figure 3.4: ***Characteristics and design stress-strain curves for steels (IS 456: 2000)***

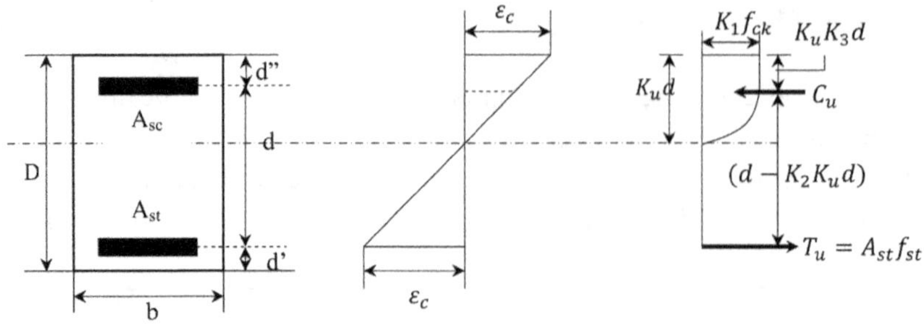

(a) Section (b) Strain (c) Stress

Figure 3.5: ***Stress-strain across RC beam section subjected to flexural loading***

Under compressive load, the stress-strain relationship for concrete is considered to in a parabolic form up to strain of 0.2%; later it remains constant up to 0.35%. Since the concrete is weak in tension, it is not allowed to crack beyond the permissible tensile stress taken as 0.7√fck. The elastic modulus considered is 5000√fck, where fck is concrete's compressive strength in N/mm2. Based on strain compatibility method, moment and curvature is calculated. Figure 3.6 represents distribution of stresses and strains across beam section subjected to flexural loading.

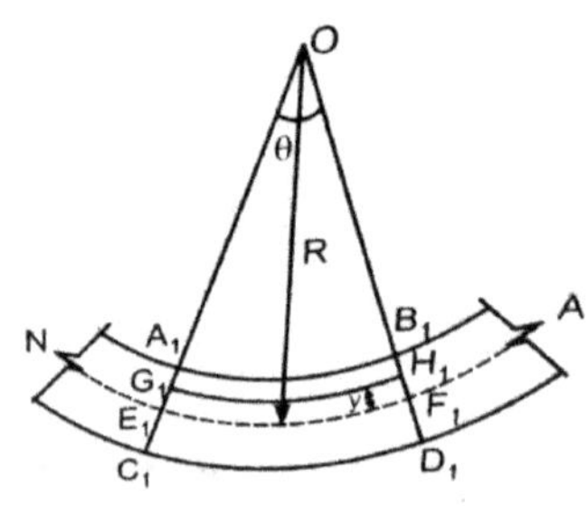

Figure 3.6: ***RC beam in bending***

Predicting moment-curvature behavior is fundamentally dependent on a thorough understanding of flexural member elements. Constrained concrete is necessary because of the lower ductility of higher-grade cements and the increased usage of these cements. The shear reinforcement in a flexure part also restricts the concrete in the compression zone. Figure 3.7 depicts the bending moment connection as follows. (Ravi Kumar et al., 2014).

$$\frac{M}{EI} = \frac{Fc}{Ec\,X} = \frac{1}{R} = \phi \tag{3.1}$$

Where, ϕ = Curvature; f_C = Compression strength of concrete at its maximum;

s_c = Concrete's maximum tensile strength; s_s = Steel at its highest tensile stress;

x = Neutral axis depth

Hence, $\phi = \frac{\varepsilon_S}{x} = \frac{\varepsilon_S}{d-x} = \frac{\varepsilon_C+\varepsilon_S}{d}$ (3.2)

Let, $\rho = \frac{Ast}{Asc}$

Modular ratio $m = \frac{Es}{Ec}$

The curvature (ϕ) is obtained as of various loads as

1. Before cracking

$f_{cr} = 0.7 \quad \sqrt{f_{ck}}$ (3.3)

Concrete resists the moment of breaking by tensing, so neglecting steel Neutral axis, N.A. = d/2.

$$M_{cr} = \frac{f_{cp} \cdot Igr}{y} \tag{3.4}$$

$I_{gr} = \frac{b\,d^3}{12};\ y = \frac{d}{2}$

$$\phi_{cr} = \frac{f_{cr}}{y} = \frac{M}{EI} \tag{3.5}$$

2. Curvature after cracking, since concrete does not take any tension N.A. depth is governedby the relationship

$$k = \sqrt{(P+P')^2 n^2 + 2n\left(p + p\frac{d'}{d}\right)} - (p - P')n \tag{3.6}$$

$$Ieff = \frac{bx^3}{3} + A_s m(D - x^2) \quad (3.7)$$

$$\phi = \frac{M_{cr}}{EI_{eff}} \quad (3.8)$$

The neutral axis at yielding is given at a distance x_u from extreme fibre, the neutral axis isfound by equation compressive force and tensile force as

$$C_u = 0.36 f_{ck} bd \quad (3.9)$$

$$T_u = A_{st} f_{st} - A_{sc} f_{sc} \quad (3.10)$$

Where in this section, the breadth b and the depth D are given., *d* is effective depth of section, *d'* is effective cover, K_u represents neutral axis factor, A_{st} is the area of tensile steel, A_{sc} is the area of compression steel f_{ck} is the characteristic cubic strength at 28 days, f_{st} is stress in steel, C_u is compressive force, T_u is tensile force.

3. Obtain neutral axis by equating the internal compressive and tensile forces. ($C_u = T_u$)

4. Compressive force as a result of concrete, yield moment is given as

$$M = A_{st} f_y (d - 0.42 x_u) + A_{sc} f_{sc} (0.42 x_u - d'') \quad (3.11)$$

$$f_y = \frac{(d-d')}{(d-x_u)} f_y$$

5. Curvature is obtained as

$$\phi_y = \frac{\varepsilon_y}{d-xu} \quad (3.12)$$

6. Obtain ultimate moment capacity of RC section by taking moment about tension steel as;

$$M_u = 0.36 f_{ck} x_{u\,max}\ b (d-o.5 x_{u\,max}) + A_{sc} f_{sc} (d - d') \quad (3.13)$$

7. Obtain curvature

$$\phi u = \frac{\varepsilon_C}{x_m} \quad (3.14)$$

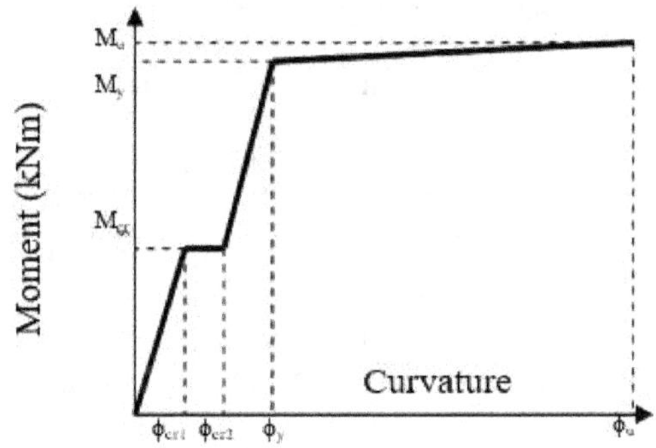

Figure 3.7: ***Moment-curvature connection for beams in an idealized form***

The idealized moment-curvature relationship for RC beam is as shown in figure 3.8.

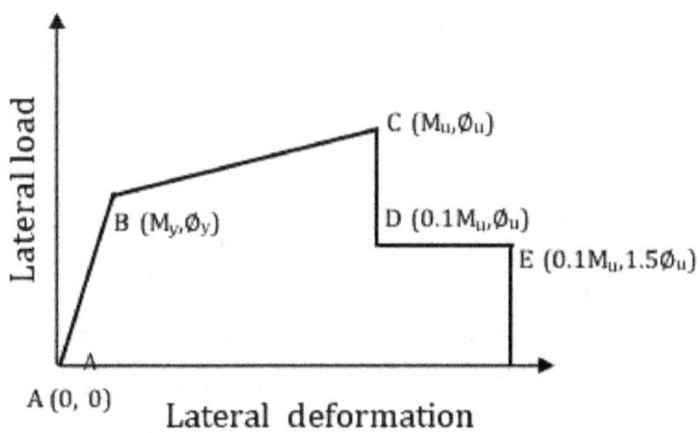

Figure 3.8: ***Input formats for multilinear moment-curvature relationships***

The moment hinges are given in the form of moment-curvature relationship in multi linear format as shown in figure 3.9. Point A is the origin with the co-ordinates of (0, 0). Point B isthe yield point and the co-ordinates are the yield moment (M_y) and yield curvature (ϕ_y). Point C is the ultimate point with the co-ordinates as ultimate moment (M_u) and ultimate curvature (ϕ_u). The other two points D and E, as required by SAP to model the residual strength of the member based on ATC-40 guidelines.

In this work the co-ordinates of point D are taken as 20% residual strength of ultimate strength. The values of M-ϕ for beams of example MRF is illustrated in Table 3.5 – 3.7.

Table 3.6: ***Values of M-ϕ for beams of example MRF***

Steps	B 1		B 2		B 3		B 4	
	M	ϕ	M	ϕ	M	ϕ	M	ϕ
At origin	0	0	0	0	0	0	0	0
Before cracking	63	0.00047	63	0.00047	63	0.00047	63	0.00047
After cracking	63	0.00091	63	0.00084	63	0.00076	63	0.00075
At yield	187.42	0.00271	212.52	0.00284	240.42	0.00291	247.11	0.00295
At ultimate	187.42	0.013	212.52	0.0130	240.42	0.0130	247.11	0.013

Table 3.7: ***Values of M-ϕ for beams of example MRF***

Steps	B5		B6		B7		B8	
	M	ϕ	M	ϕ	M	ϕ	M	ϕ
At origin	0	0	0	0	0	0	0	0
Before cracking	63	0.00047	63	0.00047	63	0.00047	63	0.00047
After cracking	63	0.00079	63	0.00078	63	0.00088	63	0.00086
At yield	230.79	0.00288	235.29	0.00291	196.09	0.00275	204.88	0.00281
At ultimate	230.79	0.013	235.29	0.013	196.09	0.013	204.88	0.013

Table 3.8: ***Values of M-ϕ for beams of example MRF***

Steps	B9		B10		B11		B12	
	M	ϕ	M	ϕ	M	ϕ	M	ϕ
At origin	0	0	0	0	0	0	0	0
Before cracking	63	0.00047	63	0.00047	63	0.00047	63	0.00047
After cracking	63	0.00109	63	0.00105	63	0.00176	63	0.0015
At yield	144.6	0.00251	153.85	0.00258	82.57	0.00231	107.31	0.00255
At ultimate	144.6	0.013	153.85	0.013	82.57	0.013	107.31	0.013

3.7.2 Axial force-moment interaction hinge (*P-M*):

Columns are designed for combine action of axial force (*P*) and bending moments (*M*) using *P-M* interaction strength envelope curve for RC members, the estimation of *P-M* interaction curve is based on limiting concrete and steel reinforcing rods are strained. as given in IS 456. In this study a simple mechanics-driven hand calculations approach to obtain design *P-M* interaction curves of rectangular RC sections are formulated based on principles of equilibrium, strain compatibility and constitutive relationships.

Design *P-M* interaction envelope is obtained as a multilinear curve made of eight salient points based on distinct normal strain distribution across the cross-section for the RC case of reinforcing steel on two sides (Aysha et al., 2015).

The column can fail when strain in concrete reaches its ultimate value s_{cu}. Similarly, to the RC members subjected to pure bending, two mode of failure occurs in case of eccentrically loaded column (i) primary tension failure initiated by yielding of the steel in tension and (ii) primary crushing of the concrete sparked the compression failure on the nearest to the direct load.

When the yielding of the tension steel and crushing of concrete in compression occurs simultaneously, the failure is called as balanced failure. The property is generated based onthe equilibrium for both forces and moments. Figure 3.9 shows a typical P-M curve generated using IS 456: 2000 recommendation.

RC sections are designed using a sequence of straight lines connecting the axial load (P) and bending moment (M) values corresponding to eight conspicuous spots identified by their unique normal strain distribution diagram as seen in the picture.

3.10. These salient points are (Aysha et al., 2015)

***Point* A:** Using the limiting compressive strain of 0.0035 as a reference, the stress in the top and bottom reinforcing steels are both identical and correspond to a strain of 0.0035, which is shown by Point A.

Point B: Represents the least compressed state of strain (bottom) edge reaches a value of 0.002, while the strain at the highly compressed (top) edge ismaintained at 0.0035; the stress in top and bottom reinforcing steels are differentalthough the stress in concrete over the entire cross section is constant (at 0.446*fck*).

Point C: Represents concrete cracking is characterized by a zero-strain state at the bottom edge of concrete, and a 0.0035-strain condition at the uppermost compressed edge. Beyond the scope of this state, a portion of concrete is subjectedto tensile strain, and therefore, cracking starts; the full design stress block of concrete in compression is mobilized over the entire depth of the cross-section.

Point D: Shows the status of the reinforcing steel's outermost layer when it is under the maximum strain. Zero stress is reached towards the bottom edge of the concrete, while 0.0035 stress is maintained at its strongly compressed (top) edge. The neutral axis depth is thus equal to the cross-effective section's depth.

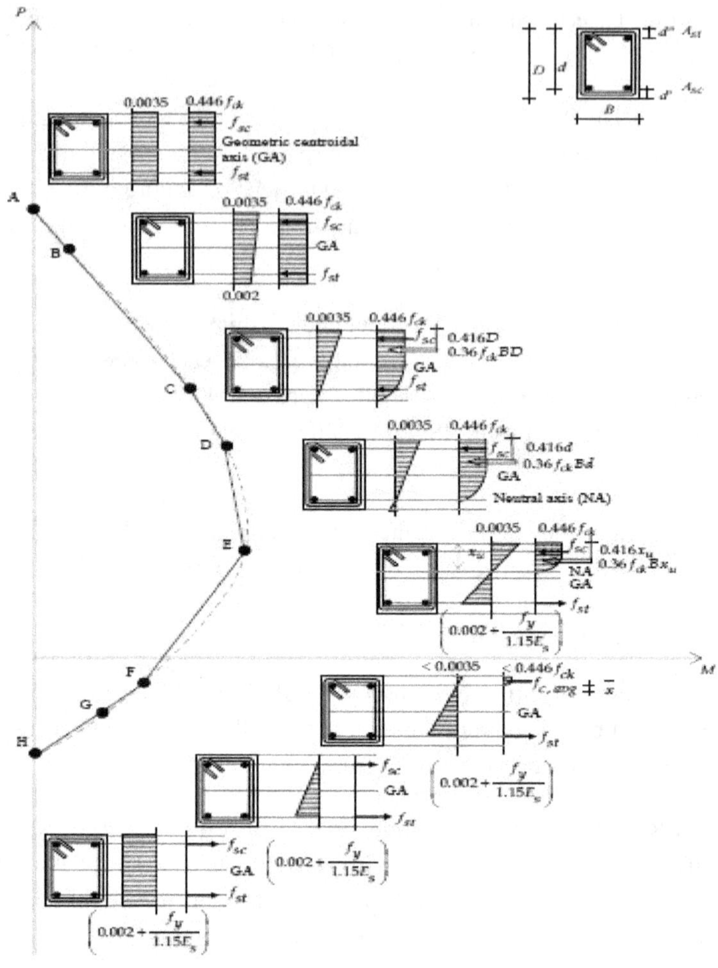

Figure 3.9: ***Axial force moment interaction curve***

Point E: It denotes a state of equilibrium – when a high degree of compression results in a balanced failure. (top) edgeof concrete in compression and the outermost layer of reinforcing steel in tension attain the limiting strain values simultaneously, namely 0.0035 and $0.002+\frac{f_y}{1{\cdot}15E_S}$ respectively; neutral axis depth can be computed directly from the strain distributiondiagram.

Point **F:** Outermost layer of reinforcing steel that is under stress has a tensile strain that is equal to or less than 0.02 $\frac{f_y}{1{\cdot}15E_S}$; while zero strain is reached at the innermost steel layer towards the compacted concrete's edge.

Point G: Represents the situation in which the Limiting strain is equivalent to strain in the outer reinforcing steel layer. of $0.002 + \frac{f_y}{1.15E_S}$; Concrete's highest edge is not subjected to any tension.

Point H: The tensile strain in the whole cross-section is equal to the limiting value under this circumstance of $0.002 + \frac{f_y}{1.15E_S}$

In order to generate a moment-rotation curve for a default plastic hinge, it was essential to make use of a stress-strain relationship that related to FEMA 356, and this was a prerequisite for doing so. This connection was implemented in software.

Equations to estimate *P* and *M* at the eight important points

The following expressions may be used to estimate P and M at each of the eight most significant locations on the design P-M interaction curve:

Point A:

$$P_A = 0.446 f_{ck} BD + (f_{sc} - 0.45 f_{ck}) A_{sc} + (f_{st} - 0.45 f_{ck}) A_{st} \quad (3.15)$$

Point B:

$$P_B = 0.446 f_{ck} BD + (f_{sc} - 0.45 f_{ck}) A_{sc} + (f_{st} - 0.45 f_{ck}) A_{st} \quad (3.16)$$

$$M_B = (f_{sc} - 0.446 f_{ck}) A_{sc} \left[\frac{D}{2} - D'\right] - (f_{sc} - 0.446 f_{ck}) A_{st} \left[\frac{D}{2} - D'\right] \quad (3.17)$$

Point C:

$$P_c = 0.36 f_{ck} BD + (f_{sc} - f_{csc}) A_{sc} + (f_{st} - f_{cst}) A_{st} \quad (3.18)$$

$$M_c = 0.36 f_{ck} BD \left[\frac{D}{2} - 0.42D\right] + (f_{sc} - 0.446 f_{ck}) A_{sc} \left[\frac{D}{2} - D'\right] - (f_{st} - 0.446 f_{ck}) A_{st} \left[\frac{D}{2} - D'\right]$$
(3.19)

Point D:

$$P_D = 0.36 f_{ck} B(D - d') + (f_{sc} - f_{csc}) A_{sc} \quad (3.20)$$

$$M_D = 0.36\, f_{ck}\, B(D - d')\left[\frac{D}{2} - \mathbf{0.416\,(D - d')}\right] + (f_{sc} - f_{csc})\, A_{sc}\left[\frac{D}{2} - \mathbf{D'}\right] \quad (3.21)$$

Point E:

$$P_E = 0.36 f_{ck} B x_u + (f_{sc} - f_{csc}) A_{sc} - f_{sc} A_{st} \quad (3.22)$$

$$M_E = 0.36\, f_{ck}\, Bx_u\left[\frac{D}{2} - 0.416Xu\right] + (f_{sc} - f_{csc})\, A_{sc}\left[\frac{D}{2} - D'\right] + f_{st}\, A_{st}\left[\frac{D}{2} - D'\right] \quad (3.23)$$

Point F:

$$P_F = f_{c,avg}\, Bd'' - f_{st}\, A_{st} \quad (3.24)$$

$$M_F == f_{c,avg}\, Bd''\left[\frac{D}{2} - (\mathbf{d''} - \bar{\mathbf{x}})\right] + f_{st}\, A_{st}\left[\frac{D}{2} - \mathbf{d'}\right] \quad (3.25)$$

Point G:

$$P_G = -f_{sc}\, A_{sc} - f_{st}\, A_{st} \quad (3.26)$$

$$M_G = -f_{sc}\, A_{sc}\left[\frac{D}{2} - d''\right] + f_{st}\, A_{st}\left[\frac{D}{2} - d'\right] \quad (3.25)$$

Point H:

$$P_H = -f_{sc}\, A_{sc} - f_{st}\, A_{st} \quad (3.26)$$

$$M_G = 0 \quad (3.27)$$

Point F:

These idioms refer to compressive loads as positive, whereas tensile loads are regarded negative in this context.

These are the predicted values for concrete's average compressive stress fc,avg, concrete's innermost reinforcing steel's compressive stress fcsc, and the distance $\bar{x}$ from the crushed edge to the concrete's resulting compressive force:

$$f_{c,avg} = \begin{cases} 0.45 * f_{ck}\left[\left(\frac{\varepsilon_c}{0.002}\right) - \frac{1}{3}\left(\frac{\varepsilon_c}{0.002}\right)^2\right] & 0 \le \varepsilon_c \le 0.002 \\ 0.45 * f_{ck}\left[1 - \frac{1}{3}\left(\frac{0.002}{\varepsilon_c}\right)\right] & 0.002 \le \varepsilon_c \le 0.0035 \end{cases} \quad (3.30)$$

$$f_{csc} = \begin{cases} 0.45 f_{ck}\left[2\left(\frac{\varepsilon_c}{0.002}\right) - \left(\frac{\varepsilon_c}{0.002}\right)^2\right] & 0 \le \varepsilon_c \le 0.002 \\ 0.45 f_{ck} & 0.002 \le \varepsilon_c \le 0.0035 \end{cases} \quad (3.31)$$

$$\bar{x} = \begin{cases} \dfrac{\left[\frac{2}{3} - \frac{1}{4}\left(\frac{\varepsilon_c}{0.002}\right)\right]}{\left[1 - \frac{1}{3}\left(\frac{\varepsilon_c}{0.002}\right)^2\right]} x_u & 0 \le \varepsilon_c \le 0.002 \\ \dfrac{\left[\frac{1}{2} - \frac{1}{12}\left(\frac{0.002}{\varepsilon_c}\right)^2\right]}{\left[1 - \frac{1}{3}\left(\frac{0.002}{\varepsilon_c}\right)\right]} x_u & 0.002 \le \varepsilon_c \le 0.0035 \end{cases} \quad (3.32)$$

In the above formulas, *Fck* is the cube strength of concrete, *fcs* for the stress in reinforcing steel, *fcsc* for the stress at the level of reinforcing steel in compression, and f_{st} for the stress in reinforcing steel in tension, respectively, in the above equations. Reinforcing steel in tension (*Ast*) and in compression (*Asc*). *d'* is the distance between the center of the reinforcing steel and the least compressed or tensile edge; *d''* is the distance between the center and the most compressed edge; *B, D,* and *D* are the dimensions of the rectangular RC section; *P* and *M* are the design axial load and moment, respectively;

Table 3.9: ***P-M interaction values for columns of example MRF***

Point	C1		C2		C3		C4	
	P	**M**	**P**	**M**	**P**	**M**	**P**	**M**
A	3909.34	0	3666.37	0	3860.4	0	3559.81	0
B	3868.07	6.77	3634.09	5.3	3820.95	6.48	3531.47	4.65
C	2277.4	94.97	2271.43	94.87	2276.2	94.95	2268.81	94.82
D	1994.09	135.26	1991.02	134.68	1993.48	135.14	1989.68	132.42
E	368.34	264.1	494.73	240.11	393.8	259.27	550.16	229.59
F	-489.64	130.92	-363.33	106.92	-464.2	126.09	-307.93	96.39
G	-746.85	78.49	-584.19	61.39	-714.09	75.04	-512.85	53.89
H	-1159.93	0	-907.3	0	-1109.1	0	-796.51	0

The values *P-M* for columns of example MRF is illustrated in Table 3.8 - 3.10. The location of beam and columns corresponding to evaluation of *M-ϕ* curve and *P-M* interaction curve is shown in figure 3.11. Figure 3.12 shows *M-ϕ* curve and *P-*

M interaction curve of example MRF.

Table 3.10: ***P-M interaction values for columns of example MRF***

Point	C5		C6		C7		C8	
	P	M	P	M	P	M	P	M
A	3800	0	3525.44	0	3705.59	0	3559.46	0
B	3762.78	6.11	3498.38	4.44	3671.86	5.54	3531.14	4.65
C	2274.71	94.93	2267.97	94.81	2272.4	94.89	2268.81	94.82
D	1992.71	135	1989.24	134.34	1991.52	134.77	1989.67	134.42
E	425.22	253.31	568.03	226.2	472.33	243.99	550.34	229.56
F	-432.8	120.12	-290.06	93	-383.72	110.79	-307.75	96.36
G	-673.66	70.79	-489.85	51.48	-610.45	64.15	-512.62	53.87
H	-1046.25	0	-760.78	0	-948.08	0	-796.15	0

Table 3.11: ***P-M interaction values for columns of example MRF***

Point	C9		C10		C11		C12	
	P	M	P	M	P	M	P	M
A	3555.64	0	3558.76	0	3487.96	0	3727.8	0
B	3527.46	4.63	3530.47	4.64	3462.28	4.21	3693.26	5.67
C	2268.71	94.82	2268.79	94.82	2267.05	94.79	2272.94	94.9
D	1989.63	134.14	1989.66	134.42	1988.77	134.25	1991.8	134.82
E	552.33	229.18	550.7	229.49	587.53	222.5	462.77	246.18
F	-305.76	95.98	-307.39	96.29	-270.57	89.3	-395.27	112.99
G	-510.06	53.6	-512.15	53.82	-464.75	48.84	-625.32	65.71
H	-792.18	0	-795.42	0	-721.8	0	-971.18	0

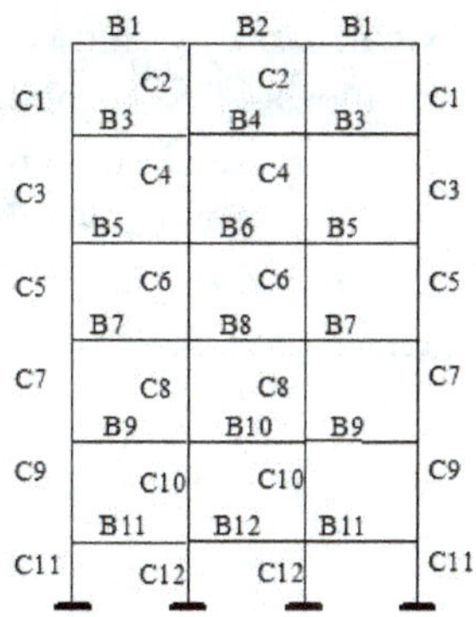

Figure 3.10: ***Location of beam and columns of example MRF***

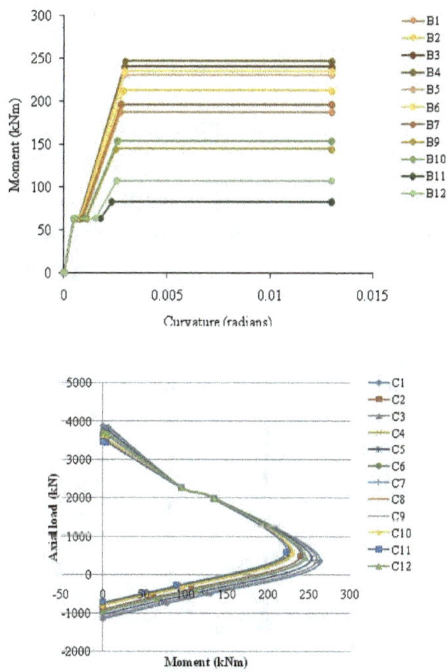

Figure 3.11: ***M-ϕ curve for beams and P-M interaction curve for columns of example MRF***

Many prior studies have demonstrated that having the ability to construct plastic hinges at both ends of a member in a frame structure with a lumped plasticity model leads to better outcomes (Mondal, 2013). Using concentrated M3 and P-M3 plastic hinges in two distinct configurations—first at both ends, then at a user-definable location—this research modeled beams and column components as nonlinear frame elements Figure 3.13 illustrates how the user-defined position was derived from previous research (Zameeruddin and Sangle, 2017c; Park and Paulay, 1975). From the basic equation, 3.32-3.37 is derived for plastic hinge lengths as follows: (Park and Paulay, 1975).

$$l_p = 0.5 \text{ H} \tag{3.33}$$

$$l_p = 0.5 \times (600) = 300 \text{ mm} \tag{3.34}$$

$$l_1 = \frac{lp}{2} = 150 \text{ mm} \tag{3.35}$$

$$l_2 = \text{H}_{\text{beam}} - \frac{lp}{2} = 450 \text{ mm} \tag{3.36}$$

$$l_3 = \text{H}_{\text{column}} - \frac{lp}{2} = 100 \text{ mm} \tag{3.37}$$

The FEMA 356 rules for modeling parameters and approval criteria have been accepted. Figure 3.13 illustrates the acceptance requirements for the final rotation capacity, which are denoted IO, LS, and CP. Table 3.11-3.13 lists the parameters and numerical acceptance requirements for the models. Sectional area, rebar percentage in tension and compression, design shear strength, as well as axial stresses all have a role.

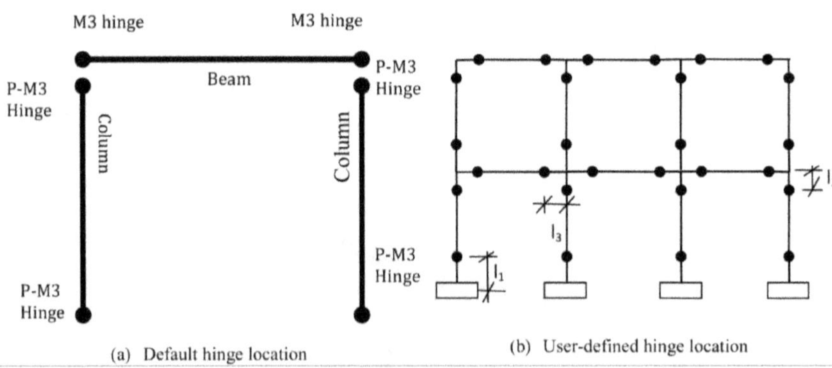

Figure 3.12: ***Location of plastic hinge***

Table 3.12: ***Plastic rotation limits for RC beams controlled by flexure*** *(FEMA 356, 2000)*

Conditions			Modeling Parameters			Acceptance Criteria				
			Plastic rotation angle (radians)		Residual strength ratio	Plastic rotation angle (radians)				
						Performance level				
						IO	Component type			
$\frac{\rho-\rho'}{\rho_{bal}}$	Trans. Reinf.	$\frac{V}{b_w d\sqrt{f_c'}}$					Primary		Secondary	
			A	b	c		LS	CP	LS	CP
≤ 0.0	C	≤ 3	0.025	0.05	0.2	0.010	0.02	0.025	0.02	0.05
≤ 0.0	C	≥ 6	0.020	0.04	0.2	0.005	0.01	0.020	0.02	0.04
≤ 0.5	C	≤ 3	0.020	0.03	0.2	0.005	0.01	0.020	0.02	0.03
≤ 0.5	C	≥ 6	0.015	0.02	0.2	0.005	0.005	0.015	0.015	0.02

For ductile details, C indicates that the transverse reinforcement satisfies the requirements

Table 3.13: ***The plastic rotation limitations for flexure-controlled RC columns are shown*** (FEMA 356, 2000)

Conditions			Modeling Parameters			Acceptance Criteria				
			Plastic rotation angle (radians)		Residual strength ratio	Plastic rotation angle (radians)				
						Performance level				
						IO	Component type			
$\frac{\rho-\rho'}{\rho_{bal}}$	Trans. Reinf.	$\frac{V}{b_w d\sqrt{f_c'}}$					Primary		Secondary	
			a	b	c		LS	CP	LS	CP
≤ 0.1	C	≤ 3	0.02	0.03	0.2	0.005	0.015	0.02	0.02	0.03
≤ 0.1	C	≥ 6	0.016	0,024	0.2	0.005	0.012	0.016	0.016	0.024
≥ 0.4	C	≤ 3	0.015	0.025	0.2	0.003	0.012	0.015	0.018	0.025
≥ 0.4	C	≥ 6	0.012	0.020	0.2	0.003	0.010	0.012	0.013	0.02

For ductile details, C indicates that the transverse reinforcement satisfies the requirements

Table 3.14: ***(Vertical components) Performance and damage levels for structures*** *(FEMA 356)*

	Structural performance level			
Maximum drift	Immediate Occupancy (IO)	Life safety (LS)	Collapse Prevention (CP)	Structural Stability
	0.01	0.02	0.04	0.33 (V_i/P_i)

Many variables are involved in this equation, including *P* (the intended load) and the shear force, *Fc* (the concrete compressive strength), B_w (the beam's width), *D* (the effective depth of the flexural component), and C (the compressive strength) (confirmation that this is the plastic zone).

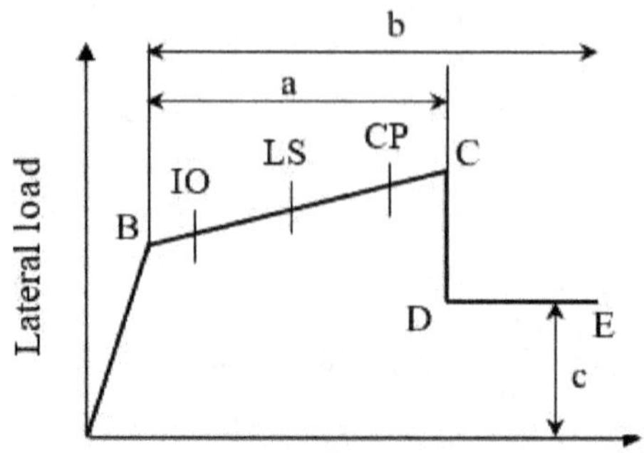

Figure 3.13: ***Assignment of hinges and various performance levels for an example MRF***

3.8 Nonlinear Static Pushover Analysis

SAP 2000 V 17.0 was used to perform a nonlinear static pushover study on the sample R. C. Structure (displacement-controlled). The analysis was split into two parts for the following reasons: Gravitational and lateral loads are the two most common forms of loads. Upper and lower limits for pushover have been suggested using lateral load patterns in literature (Ghobarah, 2000). In this study, three kinds of lateral load patterns were examined: According to figure 3.14.

4 percent of the building's overall height is the goal displacement for construction (ATC- 40). In nonlinear analysis the P-Delta geometric-nonlinearity parameter were used. The various lateral load pattern adopted in analysis are

I) Push Gravity (Push G): Dead load + 0.5 x live load
II) Push 1: IS 1893 lateral load pattern
III) Push 2: Pattern of uniform lateral load distribution
IV) Push 3: Elastic first mode load pattern in the lateral direction

3.8.1 IS 1893 load pattern: Push 1

The lateral load pattern required by IS 1893 was applied to the MRF, and it was calculated using the following expressions:

$$V_b = A_h\ Wi \tag{3.38}$$

$$A_h = \frac{z}{2}\frac{I}{R}\frac{sa}{g} \tag{3.39}$$

$$Fi = V_b\frac{w_i\ hi}{\sum w_i hi} \tag{3.40}$$

In equation 3.38 - 3.40, V_b design base shear, W_i is design seismic load (DL+0.5LL), Z is zone factor equal to 0.36, *I* represent importance factor of structure.

3.8.2 Uniform lateral load pattern: Push 2

In Push 2, As the story's mass increases, the lateral force exerted on it increases, as well:

$$F_i = V_b\left[\frac{m_i}{\Sigma m_i}\right] \tag{3.41}$$

Where F_i is the equivalent lateral force on the ith floor and m_i is the mass at the ith floor of thebuilding.

3.8.3 Elastic first-mode lateral pattern: Push 3

When it comes to vibration, the displacement mode form is closely associated with the first-mode load pattern. Stories have side forces that are proportional to the product of their elastic first-mode amplitude and their mass:

$$F_i = V_b\,\frac{M_i\phi_i}{\Sigma M_j\phi_i^2} \tag{3.42}$$

Table 3.15 illustrate the results of the modal analysis and the lateral loads, respectively.

Gravity loads and focused loads from secondary beams were used in Stage I as distributed element loads based on the yield line hypothesis. In a single step, the complete gravity load was analysed (i.e., force-control). During this examination, the structure's state was preserved and recalled in Stage II.

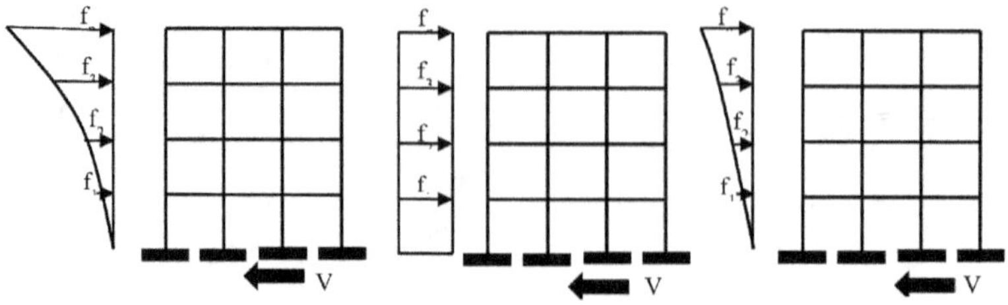

(a) IS 1893:2002 load pattern (Push 1 load case) (b) Uniform load pattern (Push 2 load case) (c) Elastic mode-1 pattern (Push 3 load case)

Figure 3.14.: ***Different load patterns used for POA***

Table 3.15: ***The example MRF's modal analysis and lateral load profile results are shown in the following sections.***

Story level	Story height (m)	Story weight (kN)	$W_i h_i^2$	$Q_i = \frac{W_i h_i^2}{\sum W_i h_i^2}$ (kN)	Obtained from SAP 2000	Loading diagram
Roof	29.1	580.23	491340.33	77.61	78.55	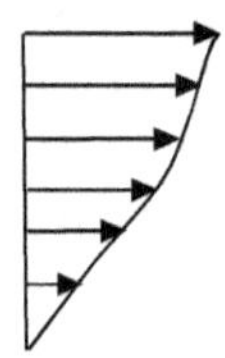
5th floor	24.1	747.20	433981.23	68.55	69.38	
4th floor	19.1	747.20	272586.03	43.06	43.58	
3rd floor	14.1	747.20	148550.83	23.46	23.74	
2nd floor	9.1	747.20	61875.63	9.77	9.90	
1st floor	4.1	735.95	12371.32	1.95	1.98	
Total		**4304.98**	**1420705.38**	**224.41**	**227.12**	

During the second stage, lateral stresses were continuously applied in small increments. A collection of lateral load patterns was chosen since POA's lateral force profile effects the structural response. A uniform load pattern, as defined in the codes or an elastic mode-1 pattern, constituted one of these sets, and upper and lower limit values for inertia loads were predicted from it. The second set included a primarily lateral load pattern (Ghobarah, 2000). For the most part, the following types of lateral load patterns were used in this investigation:

A lateral load pattern according to IS 1893 and a lateral load pattern according to elastic first mode are shown. The impacts of P- geometric nonlinearity were taken into account for each load combination employed in the investigation.

When an earthquake occurs, it has the ability to do the most damages. Because earthquake forces are irregular and non-systematic, the engineering instruments used to analyse buildings under the influence of these forces must be at their best. When it comes to seismic design, the near field ground motion (acceleration) is being replaced with performance-based design. Analyse earthquake forces to determine the structure's true behaviour and recognize that damage is predicted, but it should be coordinated. POA, which is an iterative approach, will be examined as an alternative to conventional analytic methods in this study. Low, medium and high-rise RC structures are subjected to lateral pressures with a variety of height-wise allocations until the present performance level (Purposed displacement) is achieved. The goal of performance-based seismic engineering (PBSE) is to develop structures that can withstand earthquakes. These endeavor's need a well-coordinated effort by specialists from a wide range of disciplines to achieve authenticity. Performance-based design's tremendous return is not generally accessible, which makes PBSD unique and more difficult to implement. Other than massive developments of comparable buildings, each building created using this approach is essentially unique and the expertise gained cannot immediately be transferred to constructing structures of other sorts, sizes, and performance goals. To yet, the use of Prescriptive Code Design (PBSD) has not been an economically viable alternative. PBSD is becoming more attractive to building and structural developers and engineers in seismic zones because of recent advancements in seismic hazard assessment, PBSE techniques, experimental facilities, and computer applications. In only a few short years, the PBSD technique will replace all other approaches to the design and delivery of earthquake-resistant structures. In order to make effective and efficient use of PBSD, one must be aware of the uncertainties that are present in both the structural performance estimates and the seismic hazard calculations.

3.9 Effect of seismic Loads on tall buildings

According to our perspective, increasing the height of high-rise structures has an uncertain influence on their structural stability since it increases the building's mass and height of its center of gravity. As a result, the shear force generated by seismic stresses is increased. High-rise structures are subject to earthquake loading, a sort of lateral dynamic excitation. Conceptual design, preliminary design, and optimization for gravity and lateral loads are all part of the process of designing high-rise structures. As the height of a building grows, the amount of structural material needed to withstand lateral stresses, particularly wind and earthquake loads, dramatically increases, making it more vulnerable to these loads. Since these loads may cause structural and non-structural earthquake damage, high-rise structures should be safeguarded from them.

Seismic load on a building is equal to the total of the loads on each of its levels.

Full dead load plus suitable imposed load is what constitutes each floor's seismic weight, which is the weight of each floor divided by its dead load multiplied by the appropriate amount of applied load. In this case, it comprises the weight of the permanent walls, the permanent equipment, a portion of the live load, and so on Each level above and below a storeys columns and walls must bear an equal share of the storeys weight in seismic calculations. The ground movements (accelerations) that a structure experiences as a consequence of earthquakes are likewise dynamic or variable in character, and in point of fact, they change course in a manner that is extremely haphazard. The magnitude of an earthquake force is in part determined by the magnitude of the earthquake itself, the distance from the epicenter, the local ground conditions that may amplify or dampen ground shaking, the weight (or mass) of the structure, as well as the type of structural system and its capacity to withstand abusive cyclic loading. Other factors that can play a role in determining the magnitude of an earthquake force include the type of structural system and its capacity to withstand abusive cyclic loading. The amount of lateral force that a structure is subjected to during an earthquake is exactly proportional to both the ground acceleration at the building site as well as the overall mass of the building. This relationship holds true regardless of the magnitude of the earthquake (i.e., a doubling in ground motion acceleration or building mass will double the load).

The development of skyscrapers and other tall structures is a hallmark of contemporary cities since it allows for the concentration of residential units, banks, marketplaces, exhibits, and other amenities on tiny plots of land or in a single building in the absence of available construction sites. The characteristics of a high-rise building (height, mass, high ground pressures and bottom pressures, oscillations, and difficulties evacuating in an emergency) define the hazards for those who live in the high-rise structure and those who live nearby. Factors such as terrorist attacks, wind and seismic loads, and unrestricted building operations all have a significant impact on the utilization of buildings. The horizontal component of a high-rise structure is vulnerable to earthquakes, which is one of its operational aspects. High-rise structures become more vulnerable as they grow in height and as their bulk is decreased via the use of new materials and the intelligent use of those materials' load bearing capabilities. Stiffness and frequency oscillations are reduced as a consequence of the building's reduced bulk. An increase in structural rigidity makes wind resistance greater, but it also increases the structural susceptibility to a seismic event. Modifying the cross-sectional moment of inertia of a structure is one way to alter its rigidity.

However, it is evident that the building's height is the most important factor in determining its resistance to wind and seismic loads, as well as its interior volume (area), future architectural face, and the validity of its position amid other smaller structures. The higher the building, the more susceptible it is to buckling in the wind. Increased height has an equivocal influence on the stability of high-rise structures when seismic stresses are involved, since it increases the mass of the building and raises the center of gravity. As a result, the shear force increases by a factor of three when subjected to seismic pressures. When it comes to towering structures, on the other hand, they are more flexible and better able to withstand the acceleration of the bottom. Consequently, it is essential to examine the influence of building height on its seismic resistance.

When it comes to high-rise structures and wind loads, there are several publications accessible. The seismically vulnerable high-rise structures, on the other hand, have received little attention. Numerous locations have gone years or decades without any earthquake activity. Are high-rise buildings essential in seismically active areas, given that earthquake damage to supporting structures may be as severe as damage from a fire or a terrorist act? Another problem with simulating the behaviour of a high-rise structure during an earthquake is that it is unpredictable in terms of length, force, direction, and other factors. It's also worth noting that high-rise building seismic safety systems have gathered data on wind speed and direction from meteorological measurements in this region. The topics of seismic isolation and seismic dumping are covered. When it comes to protecting buildings against earthquakes, it has been shown that bottom-mounted and other height-mountable seismic isolation systems are more effective than seism stable structures (columns, walls, frames). Only a few studies have been done on the stiffness of high-rise buildings, and they should be noted. For high-rise building passive earthquake compensation needs, a kinematic earthquake dumper system is offered by one's own design, as is the study of systems of passive and active earthquake safeguards. Roller friction and dumpers with certain stiffness and damping parameters are included in the latter. Both a model of a sliding zone with nonlinear damping and a study of basement compliance to reinforced concrete buildings under seismic impact are used to isolate the high-rise building in the event of an earthquake.

3.10 Present state of seismic analysis

Understanding structural analysis in depth is required for both the design and evaluation of seismic systems. More than a century ago, seismic regulations mandated the use of static analysis with lateral stresses equal to about 10 percent of the structure's total weight. The majority of seismic codes around the world have, for quite some time, incorporated this size in the definition of seismic loads. Over the course of history, advancements in structural dynamics and nonlinear response have been implemented, making it possible for increasingly complicated analytical procedures to be carried out. In the not too distant future, it's possible that methods including explicit probabilistic considerations will be used. This work presents a study of seismic provisions as they apply to analysis, together with a discussion of the current situation and potential future revisions. Also included in this work is a discussion of the current posture.

Probabilistic approaches to seismic performance evaluation are suitable because of the high level of uncertainty associated with the ground motion and structural modelling. Probabilistic approaches, on the other hand, are unfamiliar to most engineers, and they are reluctant to employ them. Another issue is that many in the scientific community are skeptical about explicit probabilistic methodologies other than those used in seismic hazard assessments. It appears from a quick poll of colleagues from around the world that almost all of them doubt that their countries' building codes will adopt an explicit probabilistic approach any time soon. The only country that has done so explicitly is the United States, where it has already been done in ASCE 7 but has been applied sparingly in actual construction. Codes for critical infrastructure should include an explicit quantitative evaluation of risk. But in the long run, it is impossible to totally forgo quantitative risk analysis. Risk-based design and evaluation will be required sooner or later in the most developed four countries with considerable seismicity, at the very least to improve the calibration of different safety factors and force reduction factors. Designers, building owners, and other stakeholders might benefit from information on the likelihood of earthquakes. When it comes to seismic construction standards, the application of explicit probabilistic techniques is still a long way off. Though there is some time before implementation, this should be seen as a chance to learn about and practice the methods before they are put into practice. There must be trustworthy input data and substantially simplified techniques that are known to engineers and need just a little amount of extra work and ability before quantitative risk assessment can be included in the codes. The educational value of include reliability-based content in seismic codes cannot be overstated. An informative annex to EC8 was recently drafted in Europe and is described in more detail in the following section as the first step. It was in 2012 that the Applied Technology Council (ATC) created a technique for assessing a building's seismic performance (FEMA P-58, 2012). Between 1997 and 2010, PEER researchers established a framework for performance-based seismic engineering that was used as the foundation for the technique used in this study. In the dedication of the FEMA P-58 report, A. Cornell and H. Krawinkler are recognized as the driving factors behind this framework's creation. Performance-based seismic design is the intended usage for these methods.

It is applicable for use in the assessment of both newly constructed and already existing structures. We employ a probabilistic approach, and the inherent uncertainty of the situation is taken into explicit consideration. Performance is evaluated based on the number of human casualties (deaths and severe injuries), direct economic losses (costs of repairing or replacing damaged buildings), and indirect economic losses (repair delay and dangerous placarding) that are the direct and indirect results of earthquake shaking. The implementation of the technique is broken down in companion volumes. A PACT, which is an acronym for a computerized performance assessment calculator, was made accessible to users as an additional amenity. On the other hand, the programmers do not yet include the whole approach. In order to aid in the progressive inclusion of probabilistic considerations into practice and standards, at the very least, simplified practice-oriented approaches for evaluating seismic risk are necessary. Dolek & Co. have recently completed publication of the Simplified reliability-based verification format (CEN 2017) as an Annex to EC8, Part 1. (2017b).

3.11 Next Generation Procedures (Emerging Methodologies)

Seismic risk decision-making may be improved by using new design and assessment methodologies that are based on probabilities and attempt to account for the inherent uncertainty in seismic risk. It is possible to meet the objectives of performance-based earthquake engineering by using approaches that allow for the explicit assessment of collapse risk and distinct damage states. There are already protocols in place for this kind of thing. The process presented in Appendix F of the FEMA P-695 paper, for example, uses current research findings to offer a rigorous explicit probabilistic approach for structural analysis. As a result, a structural model that can simulate collapse is required, as are several (perhaps hundreds) nonlinear response history studies, as well as explicit consideration of a wide range of potential uncertainties. It is expected that designers adopting performance-based design show via testing and analysis that the design is capable of meeting these reliability requirements. "Subject to the permission of the authorities having jurisdiction for specific projects," study and design are required. A failure probability calculation will very definitely never be done in reality since the standard allows 5 implied demonstrations that the desired dependability can be reached. Both the standard pushover-based analysis and the pushover-based N2 method, which is used for the determination of a structure's capacity, can easily estimate the annual probability of "failure" for a structure, provided that predetermined default values for dispersions are used. The Pushover-based Risk Assessment Method is a straightforward approach that follows the Annex's guidelines and needs nothing more than a typical pushover analysis. Nonlinear response history analyses replace a few pushover tests in the PRA method, which differs from Cornell's original methodology. This change was made to simplify the process (in most cases just a single one). The Pushover-based Risk Assessment Method approach, along with Cornell's closed-form solution and the basic N2 technique, falls within the category of other simpler approaches that are subject to the same limitations.

The nonlinear static analysis methodologies presented in this article will provide better results in practice. Despite this, not all of NSP's flaws have been fixed. Improvements to inelastic seismic analysis processes have revealed a number of significant insights about how they need to be developed and improved in the future. There is a need for more study on the following: First, a nonlinear model for the degradation of strength and stiffness during cyclic and in-cycle cycles has been developed. Interaction between the Soil and the foundation nonlinear multi-degree of freedom is to intensely simplicate design.

The lateral loads are induced on structure by the seismic or wind action. These loads vary with the height of the structure. Reinforced concrete structure is designed for worst combination of loads. The worst combination may include Dead loads, Live loads, Wind loads and Seismic loads with appropriate multipliers of the limits of acceptable risk. If the combination of wind loads and seismic loads are taken into account at the same time may result in to structure with higher constructional cost to mitigate ductility demands. That's the reason the seismic design and wind design are separately addressed for tall structures.

CHAPTER 4.
ANALYSIS AND INTERPRETATION

4.1 Methodology

The response and damage states of MRFs representing low, medium, and high-rise buildings were examined. For example, the MRF contains stories as indicated in figure 4.2 of Low, Medium and High-Rise. Nonlinear static analysis was performed on the sample MRF (POA). First, second, and third generation procedures were utilized to assess example MRF's response. Analytical modelling of the sample structures was performed using SAP 2000 V 21.0. One bay, three bays, and five bays, each measuring three meters in width, are shown in the MRF example in three different levels (figure 4.2): low rise, medium rise, and high rise. The example MRF was designed using IS 456:2000, IS 1893:2002 (Part 1), and IS 13920:1996. There were lateral stresses on the frame. The MRF may be found in seismic zone V, which has a z-value of 0.36, and it has a relevance factor of 1. (soil type). Structural elements are designed using a variety of materials, as shown in Table 4.1. Figure 4.1 shows the top views of an example S3B3 MRF. Table 4.1 displays information on the Material Characteristic of RC sections. Moreover, specifically the MRF in this example has three bays and four-storey for validation. A nonlinear static analysis with displacement control was carried out. on the example MRF (POA). The example of MRF's response was assessed using the criteria outlined in the very first, second, and third-creation PBSE documents.

4.2 Formulation of Work

The example MRF has a bay width of 3m and storey height of 3m. IS 456:2000, IS 1893:2002 (part 1), and IS 13920:1996 guidelines were used to create the example of MRF. the framework was subjected to lateral forces. As an example, let's look at Medium-soil MRF is located in seismic zone V (zone factor, z = 0.36) with a significance value of one. Figure 4.1 is an example of MRF's plan and elevation, which are shown in detail.

The lateral loads are induced on structure by the seismic or wind action. These loads vary with the height of the structure. Reinforced concrete structure is designed for worst combination of loads. The worst combination may include Dead loads, Live loads, Wind loads and Seismic loads with appropriate multipliers of the limits of acceptable risk. If the combination of wind loads and seismic loads are taken into account at the same time may result in to structure with higher constructional cost to mitigate ductility demands. That's the reason the seismic design and wind design are separately addressed for tall structures.

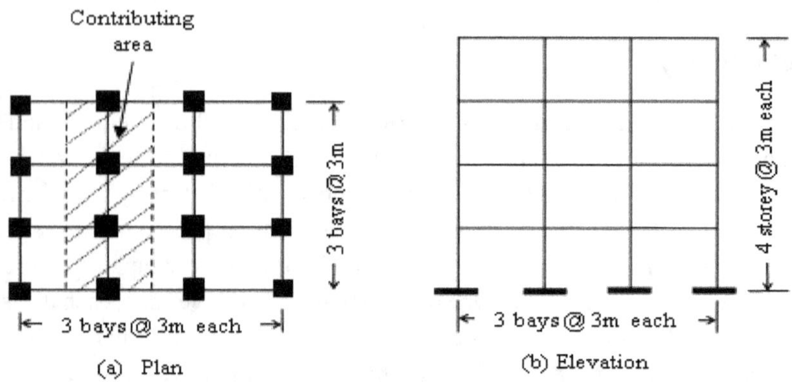

Figure 4.1: ***Typical Plan and Elevation of example S3B3 MRF***

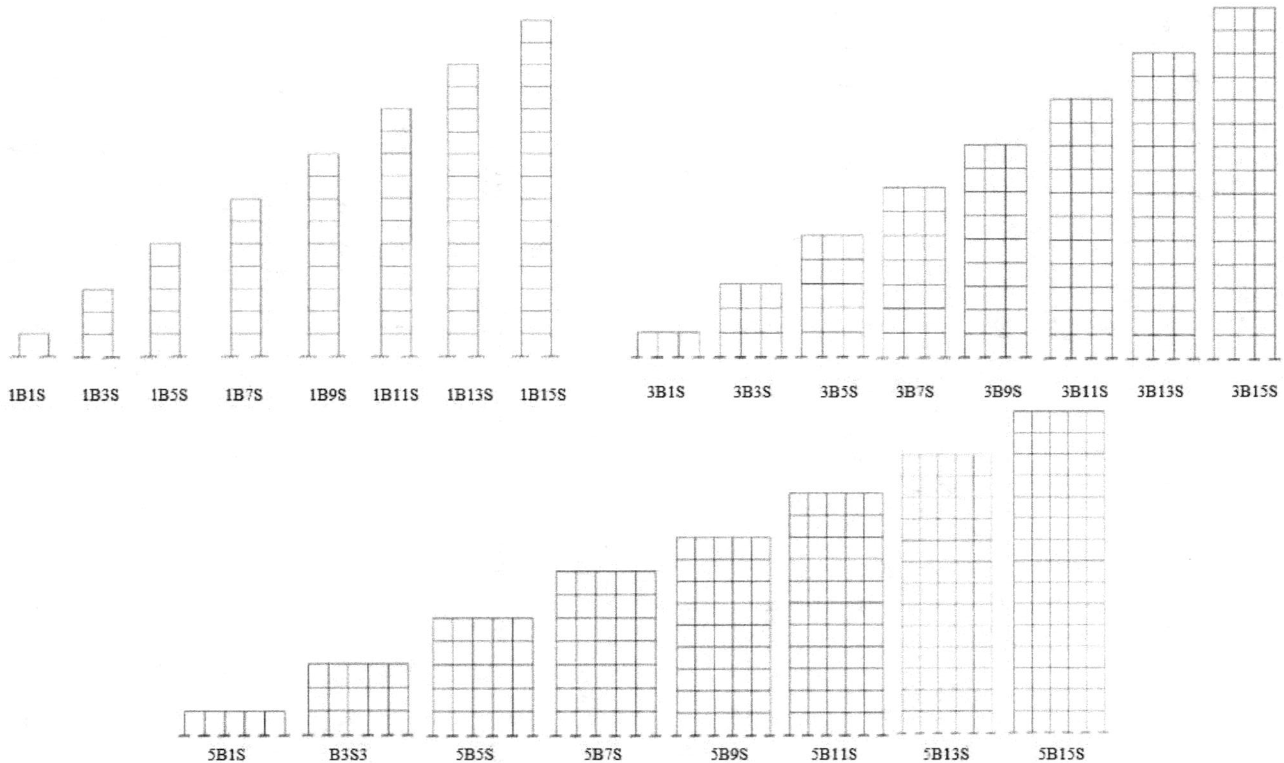

Figure 4.2 : ***Example MRFs; (a) Series-I: One bay MRFs, (b) Series-II: Three bay MRFs and (c) Series-III: Five bay MRFs***

Table 4.1: ***In an example MRF [IS 456, IS 1786], following material characteristics were taken into account***

Material Characteristic	M 25 Grade of Concrete	Steel Fe 415 grade
Weight per volume unit (kN/m^3)	25	76.97
Volume/mass (kN/m^3)	2.548	7.849
Elastic modulus of the material (kN/m^2)	25E+06	2E+08
Characteristic strength (kN/m^2)	25000 (for 28 days)	415000 (yield)
Tensile strength (Minimum) (kN/m^2)	-	485800
Yield Strength (Expected) (kN/m^2)	-	456500
Tensile Strength (Expected) (kN/m^2)	-	533500

Table 4.2: ***An overview of the intended column sections for example MRF***

Storey	**External column**		**Interior Column**	
	size (mm)	Rebar's (mm^2)	size (mm)	Rebar's (mm^2)
All floors	380 x 380	6-20 mm ϕ	450 x 450	6-20 mm ϕ

Table 4.3: ***An overview of the intended beam sections for example MRF***

Storey	**Dimension (mm)**	**Rebar's (mm^2)**								
		Bay 1			Bay 2			Bay 3		
		Top	Bot.	Top	Top	Bot.	Top	Top	Bot.	Top
1st	300 ×300	753	278	815	821	278	821	815	278	753
2nd		797	278	824	859	278	859	824	278	797
3rd		652	278	651	701	278	701	651	278	652
4th		390	345	518	532	314	532	518	345	390

The collection of lateral load patterns utilized in POA contains the following lateral load patterns: (a) lateral load pattern produced in accordance with IS 1893; (b) uniform load distribution; and (c) first mode lateral load distribution.

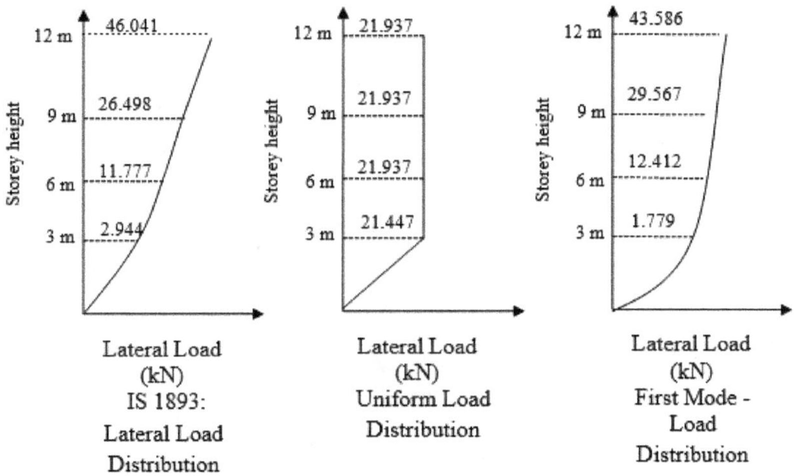

Figure 4.3: ***Application of lateral load patterns in the POA of the sample MRF***

CHAPTER 5.
CONCLUSION AND SUGGESTIONS

5.1 Assumptions

While analyzing the structures the following assumptions are being made

I) RC numbers in elastic behaviour are control by deformation and strain m - ϕ and σ - € given in Takeda model.

II) Formation of plastic hinge in each incremental performance are referred for inelastic of MRFS

III) First hinge formation is referred as operational level 'OP' which is equal to elastic deformation

IV) Collapse mechanism, sequence of plastic hinge formation is referred as in elastic behaviour of MRF

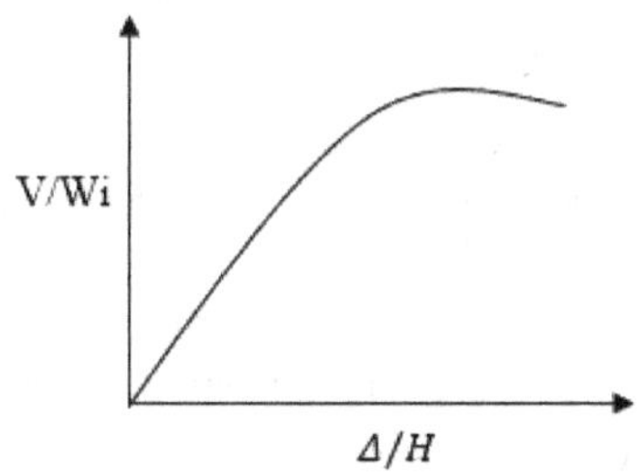

Figure 5.1: ***Pushover Analysis***

Table 5.1: ***Performance level vs Permissible limits of drift in percentage***

Sr. No.	Performance Level	Permissible limits of drift in %
1	Operational Level	0.7
2	Immediate Occupancy	1
3	Life Safety and Collapse Prevention	2.5
4	Collapse	4

5.2 Proposed Damage Indices

5.2.1 The Damage indices based on drift

Drift (conditions) restrictions are used to establish distinct degrees of performance (criteria) in PBSD (Table 5.1). It is possible to identify performance levels and collapse mechanisms using this limit, but no damage value is provided. In order to get around this problem, a drift-based damage index was put out. In Equation, the available ductility for a given performance level is used to get the damage value in DI_d. (I)

$$DI_d = \frac{d_j - d_{OP}}{d_U - d_{OP}} \qquad \text{(I)}$$

dj, dop and du are the storey displacement values at considered performance level, operational level, and permissible displacement at collapse (that is, 2.5 % H).

5.2.2 The damage indices dependent on strength.

Base shear is used in strength-based damage interactions (DIs) as a damage variable to signify a loss of strength during a POA. It is possible to utilize strength-based DI to accurately determine the behaviour of a structure during an inelastic displacement excursion. This may then lead to an objective assessment of the amount of damage that a structure has sustained. The following formula may be used to compute DIs that are based on the strength of an opponent:

For example, in POA the base shear values are defined as V (performance level), Vop (operation level), and Vmax (maximum base shear). The formula for the strength-based DIs is as follows:

$$DI_s = \frac{V_j - V_{OP}}{V_{max} - V_{OP}} \qquad \text{(II)}$$

Where; V, Vop and Vmax are the base shear values at considered performance level, operational level, and maximum base shear observed in POA.

5.2.3 The Damage index based on stiffness

The damage value is computed by matching the structure's stiffness at any defined performance level to its operational stiffness, according to the suggested damage indicator. Eq. describes the DI depending on stiffness

$$DI_k = 1 - \frac{K_J}{K_{OP}} \qquad \text{(III)}$$

Kj and Kop denote, respectively, stiffness values at the evaluated performance level and the operational level of the system.

Many more EDPs may be used to estimate damage worth, but only a few of them are mentioned here and the rest are left open for future research.

Efforts have been undertaken in this research to find out a numeric damage value for a considered performance level, using the nonlinear response obtained from the results of performance evaluation methods. The parameters used for the calibration of DI are the displacement and base shear value at identified performance levels. The yielding of plastic hinges in respective performance levels is used as performance indicators. Two performance level indicators were used to identify the example MRF performance, namely, PL1 and PL2.PL1 describes OP, IO, LS performance levels for which prescribed limits of drifts are as given in Table 5.2 whereas; PL2 corresponds to CP, C, D and E performance levels for which prescribed limits of drifts are as given in Table 5.2. The first plastic hinges formed in the A-B level, the structure is assumed to attained OP performance level. Successive fall of plastic hinges between B-C levels is used for identifying the IO, LS and CP performance levels. Table 5.2-5.4 illustrates such identifications. These DI's shows the loss in drift value (ductility), loss of strength, loss of stiffness. Many more engineering parameters may be used for damage identification, but this study has been limited for predefined DI and keeps other DI's as future scope of the study.

Table 5.2: ***Mechanism for the Collapse of Example MRF subjected to Push 1 Lateral Load case***

Step No	Displ. (m)	Base shear (kN)	A to B	B to IO	IO to LS	LS to CP	CP to C	C to D	D to E	Beyond E	Total	Remarks
3	0.014	159.66	56	0	0	0	0	0	0	0	56	Operational Level
4	0.015	167.17	51	5	0	0	0	0	0	0	56	Appearance of first hinge in B-IO Performance Level
22	0.101	253.34	29	26	1	0	0	0	0	0	56	Appearance of first hinge in IO-LS Performance Level
59	0.278	268.40	28	0	27	0	0	1	0	0	56	Appearance of first hinge in C-D Performance Level
60	0.278	210.25	28	0	25	0	0	0	3	0	56	Appearance of first hinge in D-E Performance Level

Table 5.3: ***Mechanism for the Collapse of Example MRF subjected to Push 2 Lateral Load case***

Step No	Displ. (m)	Base Shear (kN)	A to B	B to IO	IO to LS	LS to CP	CP to C	C to D	D to E	Beyond E	Total	Remarks
2	0.010	151.52	56	0	0	0	0	0	0	0	56	Operational Level
3	0.013	198.88	55	1	0	0	0	0	0	0	56	Appearance of first hinge in B-IO Performance Level
21	0.105	349.81	28	26	2	0	0	0	0	0	56	Appearance of first hinge in IO-LS Performance Level
47	0.286	368.93	28	0	25	0	0	3	0	0	56	Appearance of first hinge in C-D Performance Level
48	0.286	304.61	28	0	25	0	0	0	3	0	56	Appearance of first hinge in D-E Performance Level

The EDPs used to evaluate global damage are storey displacement, base shear, and stiffness possesses by structure at various performance levels. The proposed damage indicators are drift-based, strength-based, and stiffness-based. Proposed damage assessment is an attempt to extend the methodology proposed by Zameeruddin and Sangle (2021b) for the identification of damage zone using performance evaluation methods described in PBSD.

Table 5.4: ***Mechanism for the Collapse of Example MRF subjected to Push 3 Lateral Load case***

Step No	Displ. (m)	Base Shear (kN)	Ato B	B to IO	IO to LS	LS to CP	CP to C	C to D	D to E	Beyon dE	Tot al	Remarks
2	0.010	120.06	56	0	0	0	0	0	0	0	56	Operational Level
3	0.014	173.99	54	2	0	0	0	0	0	0	56	Appearance of first hinge in B-IO Performance Level
20	0.105	284.70	28	26	2	0	0	0	0	0	56	Appearance of first hinge in IO-LS Performance Level
55	0.286	300.22	28	0	26	0	0	2	0	0	56	Appearance of first hinge in C-D Performance Level
56	0.286	256.58	28	0	24	0	0	2	2	0	56	Appearance of first hinge in D-E Performance Level

Table 5.5: ***Nonlinear responses of example MRF***

Sr. No.	PBSE method	Push 1		Push 2		Push 2	
		Base Shear (kN)	**Displ. (m)**	**Base Shear (kN)**	**Displ. (m)**	**Base Shear (kN)**	**Displ. (m)**
1	ATC 40 (CSM)	234.48	0.064	340.28	0.050	269.72	0.059
2	FEMA 440 (CSM)	241.61	0.075	344.86	0.058	276.96	0.072
3	FEMA 356 (DCM)	251.26	0.095	346.39	0.073	283.18	0.088
4	FEMA 440 (DCM)	256.52	0.125	347.56	0.084	286.02	0.109

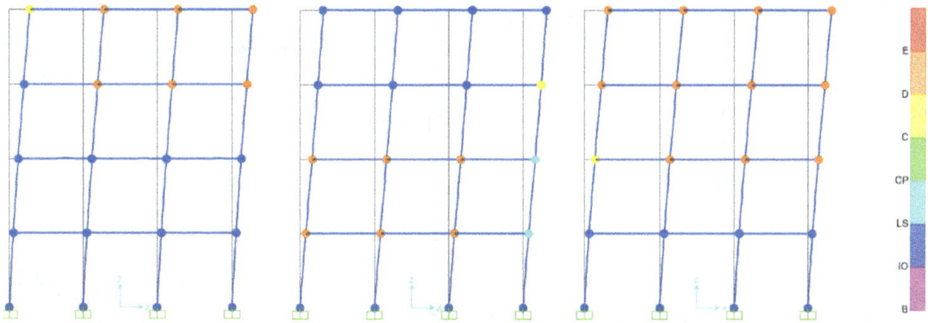

Collapse state for Push 1 load case | Collapse state for Push 2 load case | Collapse state for Push 3 load case

Figure 5.2: ***Plastic hinge mechanism of example MRF at collapse state for various lateral loads patterns***

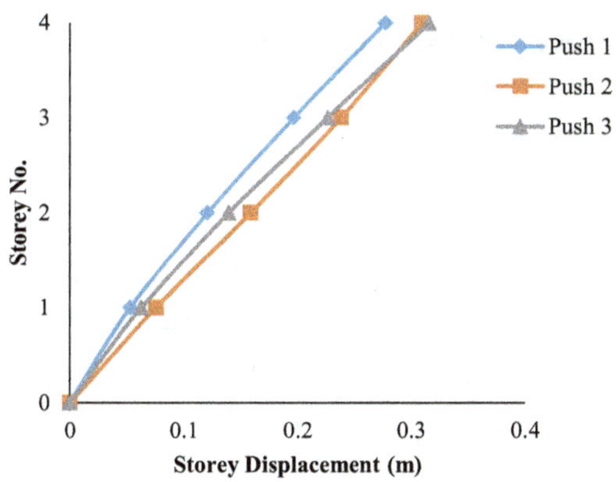

Figure 5.3: ***Storey displacement of example MRF under various lateral load cases***

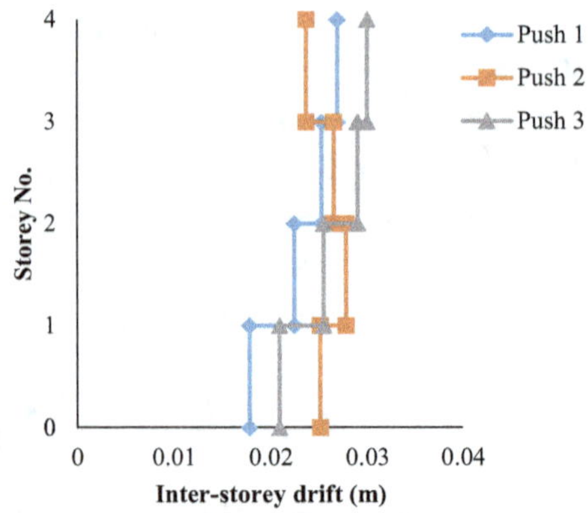

Figure 5.4: ***Inter-storey drifts of example MRF under various lateral load cases***

Table 5.6: ***Calculation of Drift-based DI values***

	Push 1				Push 2				Push 3			
Performance Level	d_{op}	d_j	d_u	DI_d	d_{op}	d_j	d_u	DI_d	d_{op}	d_j	d_u	DI_d
OP	0.014	0.014	0.30	0.000	0.010	0.010	0.30	0.000	0.010	0.010	0.30	0.00
B-IO	0.014	0.015	0.30	0.002	0.010	0.013	0.30	0.010	0.010	0.014	0.30	0.015
IO-LS	0.014	0.101	0.30	0.302	0.010	0.105	0.30	0.328	0.010	0.105	0.30	0.330
C-D	0.014	0.278	0.30	0.924	0.010	0.286	0.30	0.951	0.010	0.286	0.30	0.951
D-E	0.014	0.278	0.30	0.924	0.010	0.286	0.30	0.951	0.010	0.286	0.30	0.951

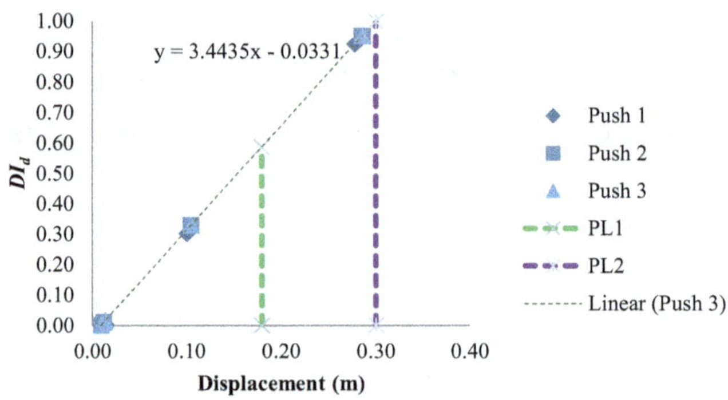

Figure 5.5: ***Drift-based DI of example MRF for PL1 and PL2***

Table 5.7: ***Calculation of Strength-based DI values***

	Push 1				Push 2				Push 3			
Performance Level	V_{op}	V_j	V_{max}	DI_s	V_{op}	V_j	V_{max}	DI_s	V_{op}	V_j	V_{max}	DI_s
OP	159.6	159.6	268.4	0.000	151.5	151.5	368.9	0.000	120	120.06	300.2	0.00
B-IO	159.6	167.1	268.4	0.069	151.5	198.8	368.9	0.218	120	173.99	300.2	0.299
IO-LS	159.6	253.3	268.4	0.861	151.5	349.8	368.9	0.912	120	284.70	300.2	0.914
C-D	159.6	268.4	268.4	1.000	151.5	368.9	368.9	1.000	120	300.22	300.2	1.000
D-E	159.6	210.2	268.4	0.465	151.5	304.6	368.9	0.704	120	256.58	300.2	0.758

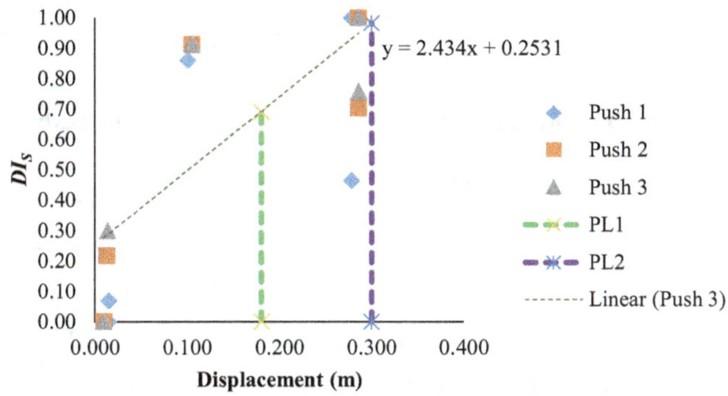

Figure 5.6: ***Strength-based DI of example MRF for PL1 and PL2***

Figure 5.6 shows the variation of DIs for various performance levels under different push lateral load cases. From the observed values, it may conclude that Push 1 and Push 2 load cases result in upper and lower bound values of example MRF. with the incremental loading on MRF fall in the stiffness of the structure is observed (as shown in figure 5.7), hence the available stiffness at a performance level can be used as a measure of damage.

The damage value of a considered performance level is evaluated by comparing the stiffness at the operational level with the available stiffness at a considered level. Table 5.8 and Figure 5.7 provide the values of DI_K obtained at various performance levels under different load cases. With the help of the trend line over distributed damage values for push 3 load case, the damage values at PL1 and PL2 are 0.65 and 1.0.

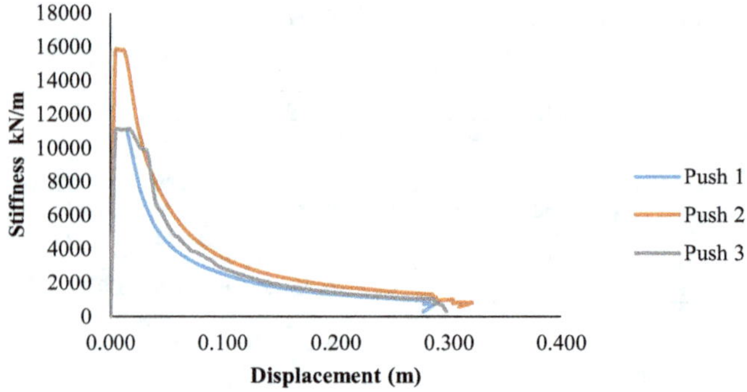

Figure 5.7: ***Variation of stiffness of example under various push load cases***

Table 5.8: ***Calculation of Stiffness-based DI values***

	Push 1			Push 2			Push 2		
Performance Level	K_{op}	K_j	DI_K	K_{op}	K_j	DI_K	K_{op}	K_j	DI_K
OP	11087.5	11087.5	0.000	15783.4	15783.4	0.000	12506.8	12506.8	0.00
B-IO	11087.5	11087.5	0.000	15783.4	15783.3	0.000	12506.8	12506.6	0.000
IO-LS	11087.5	2517.43	0.773	15783.4	3338.52	0.788	12506.8	2703.52	0.784
C-D	11087.5	964.79	0.913	15783.4	1291.54	0.918	12506.8	1050.12	0.916
D-E	11087.5	755.74	0.932	15783.4	1066.36	0.932	12506.8	897.43	0.928

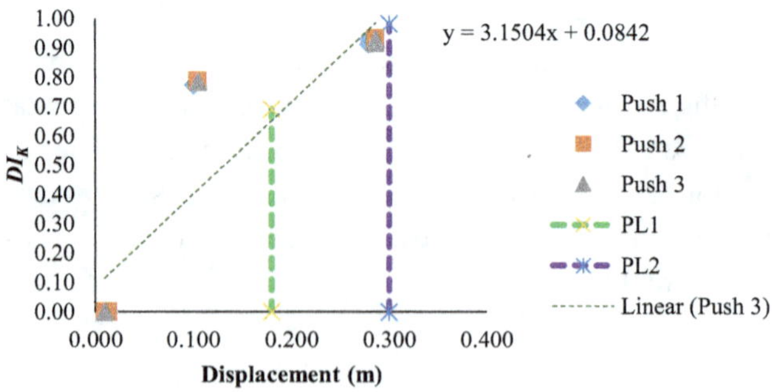

Figure 5.8: ***Stiffness-based DI of example MRF for PL1 and PL2***

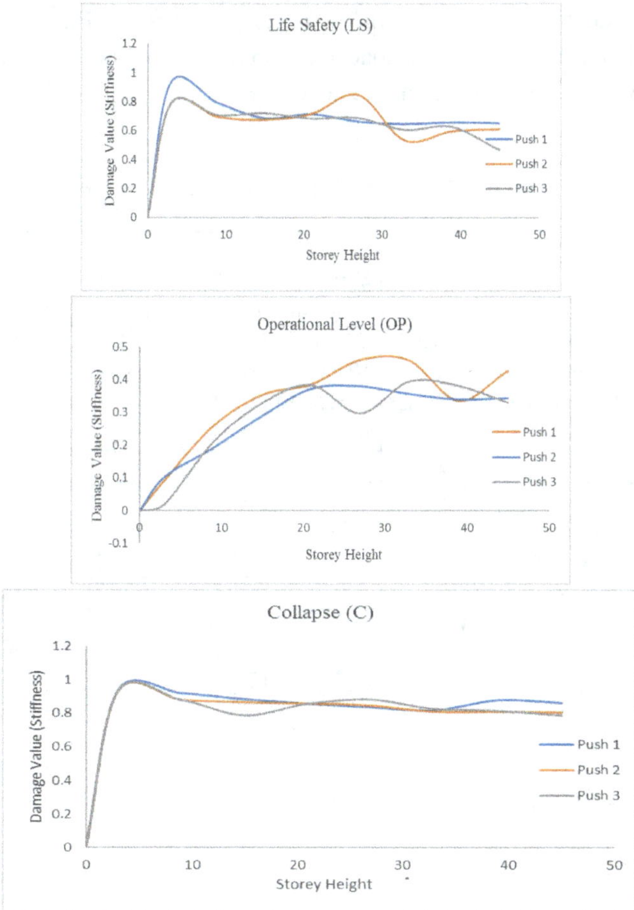

Figure 5.9: ***Damage variations in Series-II MRFs with three storey subjected to different Push Load cases.***

Its rational approach to evaluate the state of damage to a structure which may referred for valuation of vulnerability index. For Push 3, Push 1, and Push 2, results from POA reveal that MRFs have lower limits, upper bounds, and median values. It was shown that ductile failure mechanisms were seen when plastic hinge mechanisms were formed in the collapse stage (figure 5.2), whereas brittle failure mechanisms were observed in the Push 3 Load Case. When comparing the drift numbers for Push 2 and Push 3 load situations, the difference is about 10% smaller. Reinforced concrete members exposed to cyclic loading lose drift due to nonlinear yielding of higher stories, hence it is necessary to use various push loading instances to depict the whole inelastic excursion. Examples of MRFs exposed to POA for various Push load scenarios are shown in Table 5.3. Examples of MRFs with varying performance levels were used to calculate the damage values.

Table 5.9: ***Mechanism for the Collapse of Example MRFs subjected to Push 1 Load case***

Operational Level (OP)						
Series I (Low Rise)			Series II (Medium Rise)		Series III (High Rise)	
Storey Height	Damage Value (Strength)	Damage Value (Stiffness)	Damage Value (Strength)	Damage Value (Stiffness)	Damage Value (Strength)	Damage Value (Stiffness)
0	0	0	0	0	0	0
3.000	0.343	0.092	0.789	0.103	0.567	0.020
9.000	0.591	0.258	0.766	0.192	0.736	0.210
15.000	0.553	0.353	0.747	0.291	0.747	0.330
21.000	0.681	0.385	0.724	0.373	0.727	0.384
27.000	0.774	0.459	0.663	0.381	0.577	0.298
33.000	0.713	0.458	0.603	0.358	0.613	0.395
39.000	0.790	0.335	0.611	0.341	0.610	0.382
45.000	0.795	0.426	0.620	0.345	0.526	0.331

Table 5.10: ***Mechanism for the Collapse of Example MRF subjected to Push 2 Load case***

Life Safety (LS)						
Series I (Low Rise)			Series II (Medium Rise)		Series III (High Rise)	
Storey Height	Damage Value (Strength)	Damage Value (Stiffness)	Damage Value (Strength)	Damage Value (Stiffness)	Damage Value (Strength)	Damage Value (Stiffness)
0	0	0	0	0	0	0
3.000	0.991	0.934	0.926	0.789	0.926	0.788
9.000	0.916	0.793	0.870	0.693	0.859	0.708
15.000	0.910	0.687	0.840	0.674	0.871	0.721
21.000	0.865	0.715	0.854	0.714	0.865	0.685
27.000	0.864	0.664	0.900	0.846	0.787	0.689
33.000	0.856	0.648	0.928	0.532	0.812	0.607
39.000	0.950	0.658	0.808	0.596	0.819	0.628
45.000	0.992	0.653	0.811	0.610	0.876	0.466

Table 5.11: ***Mechanism for the Collapse of Example MRF subjected to Push 3 Load case***

Collapse (C)						
Series I (Low Rise)		Series II (Medium Rise)			Series III (High Rise)	
Storey Height	Damage Value (Strength)	Damage Value (Stiffness)	Damage Value (Strength)	Damage Value (Stiffness)	Damage Value (Strength)	Damage Value (Stiffness)
0	0	0	0	0	0	0
3.000	0.991	0.934	0.994	0.931	0.995	0.930
9.000	0.901	0.921	1.000	0.882	1.000	0.881
15.000	1.000	0.885	1.000	0.866	1.000	0.787
21.000	0.996	0.856	1.000	0.858	1.000	0.857
27.000	0.998	0.836	1.000	0.846	0.994	0.882
33.000	1.000	0.818	1.075	0.810	0.995	0.827
39.000	0.909	0.878	0.995	0.810	0.994	0.815
45.000	0.927	0.862	0.993	0.806	1.000	0.787

Table 5.12. ***Nonlinear responses of example MRF in reference to various PBSE methods***

Sr. No.	PBSE Method	Push 1		Push 2		Push 2	
		Base Shear (kN)	Displ. (m)	Base Shear (kN)	Displ. (m)	Base Shear (kN)	Displ. (m)
1	ATC 40 (CSM)	234.48	0.064	340.28	0.050	269.72	0.059
2	FEMA 440 (CSM)	241.61	0.075	344.86	0.058	276.96	0.072
3	FEMA 356 (DCM)	251.26	0.095	346.39	0.073	283.18	0.088
4	FEMA 440 (DCM)	256.52	0.125	347.56	0.084	286.02	0.109

It was designed to accommodate a variety of load scenarios. To scale up the damage value, researchers in this study used the nonlinear response from PBSE approaches. Various displacements and base shear values are utilized to estimate DI's. The drop-in performance levels of plastic hinges are utilized as a performance indicator. There were two performance level indicators, PL1 and PL2, used to determine the example MRF performance under various lateral load patterns.

Table 5.13: ***The performance levels, as well as the damage and drift restrictions that correspond to them*** **[Ahmed Ghobarah 2001]**

Performance	Damage state	Drift	Performance Level
Fully operational, immediate occupancy	No damage	<0.32%	**Performance Level-1 (PL-1)**
Operational, damage control, Moderate	Repairable	<0.5%	
Life in a non-harmful condition.	Irreparable	<1.5%	**Performance Level-2 (PL-2)**
Near collapse, limited safety, hazard reduce	severe	<2.5%	
Collapse		>2.5%	

PL1 refers to performance levels for which the limits of drifts are listed in Table 5.13 however, PL2 refers to performance levels for which the limits of drifts are not included. A-B level plastic hinges are believed to be in OP, and subsequent falls between B-C level plastic hinges are utilized to designate IO, LS, and CP levels. In Table 5.2-5.4, such identifications are described. The DI's reflect a decrease in drift, strength, and stiffness. It is possible to identify damage using many more engineering criteria, however this study has been confined to a specific DI and has left room for future research into additional DI's.

According to Equations. 1 to 3, the values of DI's, such as MRFs, may be found in Table 5.9-5.11. A look at the calculated values of DI d shows that the damage thresholds for P1 are lower than the allowed limits for drift (see figure 5.10 and Table 5.13). Comparing PL1 to PL2, damage levels at PL2 were within the permitted limits for drift values. Linearization results in PL1 and PL2 damage values of 0.59 and 1.0, respectively. The RC sections' balance between compressive and tensile forces, which is dependent on the steel quality, may be to blame for this behaviour. To increase the ductility of the RC sections, the identification of such a zone helps. In order to find a comprehensive answer to these issues, researchers will need to conduct several case studies. A comprehensive answer is left out of this study's scope, but the discovery of grey areas for improvement is the primary goal.

Table 5.14: ***Calculation of Drift-based DI value***

	Push 1				Push 2				Push 3			
Performance Level	d_{op}	d_j	d_u	DI_d	d_{op}	d_j	d_u	DI_d	d_{op}	d_j	d_u	DI_d
OP	0.014	0.014	0.30	0.000	0.010	0.010	0.30	0.000	0.010	0.010	0.30	0.00
B-IO	0.014	0.015	0.30	0.002	0.010	0.013	0.30	0.010	0.010	0.014	0.30	0.015
IO-LS	0.014	0.101	0.30	0.302	0.010	0.105	0.30	0.328	0.010	0.105	0.30	0.330
C-D	0.014	0.278	0.30	0.924	0.010	0.286	0.30	0.951	0.010	0.286	0.30	0.951
D-E	0.014	0.278	0.30	0.924	0.010	0.286	0.30	0.951	0.010	0.286	0.30	0.951

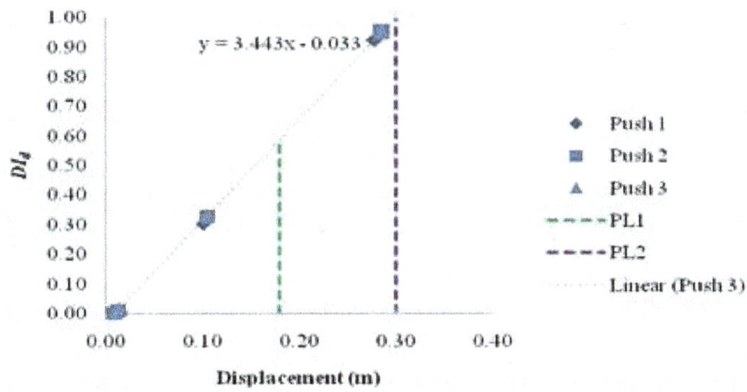

Figure 5.10: ***Drift-based DI of example MRF for PL1 and PL2***

Table 5.14 provides the values DI_s obtained from nonlinear responses resulting from POA performed on example MRF. By linearization technique applied on results for Push 3 load case damage value obtained at PL1 and PL2 is 0.69 and 0.98. figure 5.10 show the variation of DI_s for various performance levels under different push load cases. From the observed values it may conclude that Push 1 and Push 2 load cases results upper and lower bound values of example MRFs. The trends of obtained results shows that Push 1 forms upper bound envelops while Push 3 load case results in lower bound values in terms of nonlinear response.

Table 5.15: ***Calculation of Strength-based DI value***

Performance Level	Push 1				Push 2				Push 3			
	V_{op}	V_j	V_{max}	DI_s	V_{op}	V_j	V_{max}	DI_s	V_{op}	V_j	V_{max}	DI_s
OP	159.6	159.6	268.4	0.000	151.5	151.5	368.9	0.000	120	120.06	300.2	0.00
B-IO	159.6	167.1	268.4	0.069	151.5	198.8	368.9	0.218	120	173.99	300.2	0.299
IO-LS	159.6	253.3	268.4	0.861	151.5	349.8	368.9	0.912	120	284.70	300.2	0.914
C-D	159.6	268.4	268.4	1.000	151.5	368.9	368.9	1.000	120	300.22	300.2	1.000
D-E	159.6	210.2	268.4	0.465	151.5	304.6	368.9	0.704	120	256.58	300.2	0.758

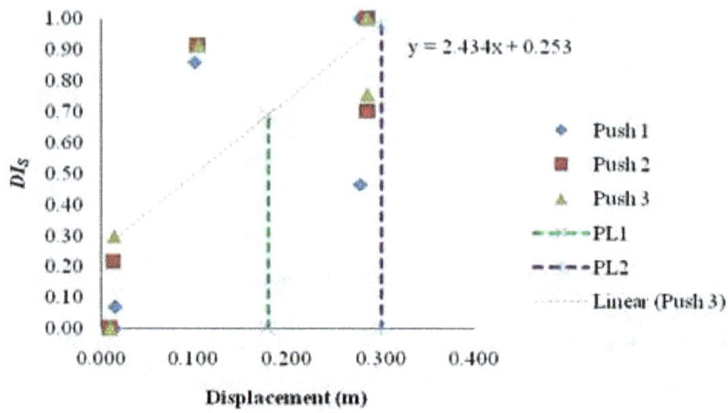

Figure 5.11: ***Strength-based DI of example MRF for different push load cases***

With the incremental loading there is a fall in the stiffness of the structure (as shown in figure 5.12), hence the available stiffness at a performance level can be used as a tool for assessing damage. The damage value of a performance level that is being analyzed is determined by contrasting the level of available stiffness at that level with the level of stiffness that is present at the operational level. Table 5.15 and figure 5.11 provide the values of DI_K obtained at various performance levels under different load cases. With the help of linearization DI_K at any value of displacement can be obtained. From the obtained relationship for push 3 load case the damage values at PL1 and PL2 are 0.65 and 1.0.

Table 5.16: ***Comparison of drift limits for various Push load cases***

PUSH 1 Load Case

Performance Level	Drift Limit in % (Permissible)	**Series 1 (Low Rise)**							
		S1B1	S3B1	S5B1	S7B1	S9B1	S11B1	S13B3	S15B1
OP	0.2	0.00	0.01	0.01	0.01	0.05	0.67	0.04	0.05
LS	1.5	0.07	0.22	0.34	0.32	0.46	1.36	0.65	0.73
C	>2.5	0.23	0.69	1.15	1.46	1.49	3.91	2.53	2.39
Performance Level	Drift Limit in % (Permissible)	**Series 2 (Medium Rise)**							
		S1B3	S3B3	S5B3	S7B3	S9B3	S11B3	S13B3	S15B3
OP	0.2	0.00	0.01	0.02	0.02	0.02	0.00	0.04	0.07
LS	1.5	0.06	0.19	0.30	0.36	0.49	0.05	0.77	0.94
C	>2.5	0.23	0.68	1.18	1.63	2.08	1.36	2.93	3.53
Performance Level	Drift Limit in % (Permissible)	**Series 3 (High Rise)**							
		S1B5	S3B5	S5B5	S7B5	S9B5	S11B5	S13B3	S15B5
OP	0.2	0.00	0.01	0.01	0.02	0.02	0.04	0.06	0.01
LS	1.5	0.06	0.19	0.28	0.39	0.43	0.53	0.57	0.64
C	>2.5	0.23	0.69	1.13	1.64	1.62	2.35	2.23	2.68

PUSH 2 Load Case

Performance Level	Drift Limit in % (Permissible)	**Series 1 (Low Rise)**							
		S1B1	S3B1	S5B1	S7B1	S9B1	S11B1	S13B3	S15B1
OP	0.2	0.00	0.01	0.01	0.01	0.04	0.57	0.03	0.05
LS	1.5	0.07	0.22	0.34	0.35	0.55	1.54	0.75	0.83
C	>2.5	0.35	0.79	1.25	1.56	1.69	3.01	2.65	2.59
Performance Level	Drift Limit in % (Permissible)	**Series 2 (Medium Rise)**							
		S1B3	S3B3	S5B3	S7B3	S9B3	S11B3	S13B3	S15B3
OP	0.2	0.00	0.01	0.02	0.02	0.02	0.00	0.03	0.06
LS	1.5	0.07	0.20	0.35	0.40	0.49	0.05	0.77	0.95
C	>2.5	0.25	0.70	1.28	1.69	2.58	1.46	2.93	3.60
Performance Level	Drift Limit in % (Permissible)	**Series 3 (High Rise)**							
		S1B5	S3B5	S5B5	S7B5	S9B5	S11B5	S13B3	S15B5
OP	0.2	0.00	0.12	0.15	0.02	0.04	0.04	0.06	0.01
LS	1.5	0.02	0.29	0.30	0.39	0.45	0.54	0.60	0.65
C	>2.5	0.25	0.70	1.17	1.69	1.64	2.39	2.29	2.60

PUSH 3 Load Case

Performance Level	Drift Limit in % (Permissible)	Series 1 (Low Rise)							
		S1B1	S3B1	S5B1	S7B1	S9B1	S11B1	S13B3	S15B1
OP	0.2	0.00	0.01	0.01	0.01	0.05	0.60	0.04	0.05
LS	1.5	0.08	0.32	0.44	0.45	0.65	1.84	0.85	0.93
C	>2.5	0.45	0.89	1.35	1.66	1.79	3.11	2.75	2.69
Performance Level	Drift Limit in % (Permissible)	Series 2 (Medium Rise)							
		S1B3	S3B3	S5B3	S7B3	S9B3	S11B3	S13B3	S15B3
OP	0.2	0.01	0.07	0.01	0.03	0.03	0.00	0.04	0.07
LS	1.5	0.05	0.09	0.55	0.50	0.59	0.05	0.87	0.95
C	>2.5	0.35	0.80	1.38	1.79	2.78	1.66	2.93	3.80
Performance Level	Drift Limit in % (Permissible)	Series 3 (High Rise)							
		S1B5	S3B5	S5B5	S7B5	S9B5	S11B5	S13B3	S15B5
OP	0.2	0.00	0.22	0.25	0.03	0.05	0.05	0.07	0.05
LS	1.5	0.02	0.29	0.30	0.59	0.55	0.56	0.70	0.85
C	>2.5	0.27	0.70	1.17	1.69	1.84	2.49	2.39	2.76

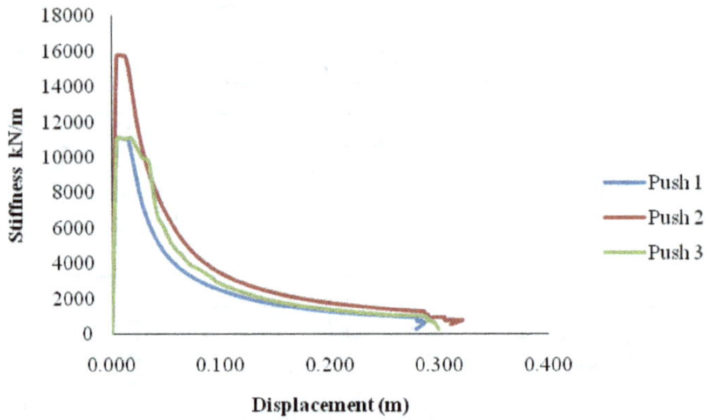

Figure 5.12: ***Loss of stiffness during POA for different load case***

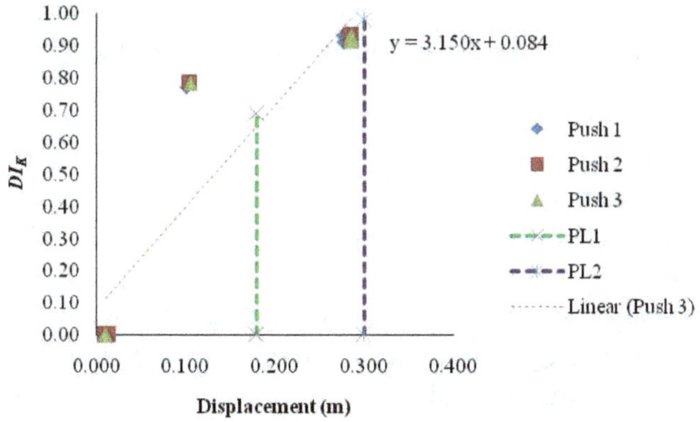

Figure 5.13: ***Stiffness-based DI of example MRF push load cases***

5.3 Conclusions & Future Scope

Its rational approach to evaluate the state of damage to a structure which may referred for evaluation of vulnerability index. Damage to the structures during a seismic event is a prime concern of structural engineers. The damage indicators available in the literature are mostly based on responses obtained from the nonlinear dynamic analysis, with few attempts wherein nonlinear static analysis has been used. Nonlinear dynamic analysis involves is a complex and tiresome process, hence is not common in practice. PBSD has provided various PBSE methods using POA for performance evaluation, but they themselves are not capable to provide any numerical value of damage sustain by structure and its components. The present study attempts to provide a possible way of integrating the damage and performance parameters for damage assessment of a structure using the responses resulted from POA performed on the structure and capable of addressing cyclic loading effects. A review of the state-of-art-of-development in the DI's formulation at various state of building performance levels has been provided. In addition, an attempt has been made to integrate the PBSE procedure with damage values by defining the DI's using EDP's resulting from POA.

The results obtained from NLSP performed on example MRF were used to find damage value in reference to the performance level of a structure. To identify the state of damage to the structure two performance level indicators were set up namely PL1 and PL2. The damage variable used in the formulation of various proposed damage indicators is capable to track the cumulative effects at every incremental displacement which overcomes the limitation of POA. With the help of the PL1 and PL2 performance levels, one can identify the zone of damage and the associated scope for the structural design optimization.

Damage to the structures during a seismic event is a prime concern of structural engineers. The damage indicators available in the literature are mostly based on responses obtained from the nonlinear dynamic analysis, with few attempts wherein nonlinear static analysis has been used. Nonlinear dynamic analysis involves is a complex and tiresome process, hence is not common in practice. PBSD has provided various PBSE methods using POA for performance evaluation, but they themselves are not capable to provide any numerical value of damage sustain by structure and its components. The present study attempts to provide a possible way of integrating the damage and performance parameters for damage assessment of a structure using the responses resulted from POA performed on the structure and capable of addressing cyclic loading effects. A review of the state-of-art-of-development in the DI's formulation at various state of building performance levels has been provided. In addition, an attempt has been made to integrate the PBSE procedure with damage values by defining the DI's using EDP's resulting from POA.

The results obtained from NLSP performed on example MRF were used to find damage value in reference to the performance level of a structure. To identify the state of damage to the structure two performance level indicators were set up namely PL1 and PL2. The damage variable used in the formulation of various proposed damage indicators is capable to track the cumulative effects at every incremental displacement which overcomes the limitation of POA. With the help of the PL1 and PL2 performance levels, one can identify the zone of damage and the associated scope for the structural design optimization.

Structural engineers are increasingly worried about earthquake-related damage to structures. In the past many of damage indicators were defined using parameters resulting from nonlinear dynamic analysis, but tiresome procedure involved has fail to make them common in practice. PBSD has given PBSE techniques employing NLSP for performance assessment of structures, which are easy for implementation and provides results nearer to nonlinear dynamic analysis hence became common in practice. But they are unable to provide any numerical value showing the damage incurred by the structure and its components. In this study, we have introduced a technique for combining the performance and damage parameters of a structure resulting in quick damage assessment using the parameters from POA. The defined damage indices have been assessed at different building performance levels to establish the relation between performance level and associated damage state.

Two performance level indicators, PL1 and PL2, were designed to measure the state of damage to the building. Using a damage variable that accounts for the cumulative displacement, stiffness and strength at each incremental lateral load helps to overcome the POA limitation to address inelastic cyclic loading effects. Now damage may be measured at any point referring the capacity curve. Performance levels PL1 and PL2 helps for structural design optimization.

Bibliography

[1]. Alpa Sheth, and Sudhir K Jain (2002). Training of teachers for capacity building towards earthquake safety in India. The Indian Concrete Journal.

[2]. Andre F, Eric L and Partrick L (1998). Seismic performance of ductile and nominally ductile reinforced concrete moment resisting frames, I. Experimental study, Canadian Journal of Civil Engineering, 25: 331-341

[3]. Arjomandi, K., Eslekanchi, H., Vafai, A., (2009). Correlation between structural performance levels and damage indexes in steel frame subjected to earthquakes. Trans. A Civil Eng. Sharif Univ. 16 (2), 147–155.

[4]. ASCE (2000). Prestandard and Commentary for the Seismic Rehabilitation of Buildings. FEMA 356 Report, prepared by the American Society of Civil Engineers for the Federal Emergency Management Agency, Washington, DC.

[5]. ASCE 41, 2006. Seismic rehabilitation of existing buildings. American society of civil engineers, Reston, Virginia. ATC 40, 1996. Seismic evaluation and retrofit of the existing concrete building. Applied Technical Council, Redwood City (CA).

[6]. ASCE/SEI 41 (2007). Seismic rehabilitation of existing building. American Society of Civil Engineers, Reston, Virginia.

[7]. ATC 40, 1996. Seismic evaluation and retrofit of existing concrete building. Applied Technical Council, Redwood City (CA).

[8]. Aysha Z. Majeed, Rupen Goswami and CVR Murty (2015). Mechanics - driven Hand Calculation Approach for Obtaining Design P-M Interaction Curves of RC Sections. Indian Concrete Journal, 89(9), pp. 59-68

[9]. Borg, R.C., Rossetto, T., (2010). Performance based seismic design and assessment methodologies: relation to damage and requirements. 8th Fib PhD Symposium in Kgs, Lyngby, Denmark.

[10]. Boroujeni, A.R.K., 2013. Evaluation of various methods of FEMA 356 compare to FEMA 440. J. Civil Eng. Constr. Technol. 4 (2), 51–55.

[11]. Chopra AK, Goel RK (2000) Evaluation of NSP to estimate seismic deformation: SDF system, Journal of Structural Engineering; 126 (4):482–90

[12]. Comartin CD, Aschhein M, Guyader A, Hamburger R, Hanson R, Holmes W, et al. (2004). A summary of FEMA-440: improvement of nonlinear static seismic analysis procedures. The 13th world conference on earthquake engineering, 2004 Vancouver, B.C., Canada Paper No. 1476

[13]. Erol K and Kunnath SK (2007). Assessment of current nonlinear static procedures for seismic evaluation of buildings. Engineering Structures, 29: 305–316

[14]. FEMA 273, 1996. NEHRP guidelines for the seismic rehabilitation of buildings. Federal Emergency Management Agency, Washington (DC).

[15]. FEMA 356, (2000). Pre-standard and commentary for seismic rehabilitation of buildings. Federal Emergency Management Agency, Washington (DC).

[16]. FEMA 440, (2005). Improvement in nonlinear static seismic analysis procedures. Federal Emergency Management Agency, Washington (DC).

[17]. FEMA 445, (2006). Next-generation performance-based seismic design guideline program for new and existing buildings. Federal Emergency Management Agency, Washington (DC).

[18]. Freeman SA (2000). Performance-based seismic engineering: past, current and future.

[19]. Freeman SA (2004). Review of the development of the capacity spectrum method. ISET Journal Earthquake Technology; 41(1): 1–13

[20]. Ghobarah, A., (2000). Seismic assessment of existing RC structures. Prog. Struct. Eng. Mater. 2, 60–71.

[21]. Ghobarah, A., (2001). Performance-based design in earthquake engineering: state of development. Eng. Struct. 23, 878–884.

[22]. Ghobarah, A., (2004). On drift limits associated with different damage levels. In proceedings of the international workshop on performance-based seismic design, concepts and implementation, Bled, Slovenia.

[23]. Ghobarah, A., Abo-Elfath, H., Biddah, A., (1999). Response-based damage assessment of structures. Earthquake. Eng. Struct. Dynamics. 28, 79–104.

[24]. Ghosh, S., Datta, D., Katakdhond, A.A., (2011). Estimation of the Park-Ang damage index for planar multi-storey frames using equivalent single-degree systems. Engineering Structures 33, 2509–2524.

[25]. Habibi, A.R., Izadpanah, M., (2013). New method for the design of reinforced concrete moment resisting frames with damage control. Sci. Iran. Trans. A 19 (2), 234– 241.

[26]. Habibi, A.R., Izadpanah, M., Yazdani, A., (2013). Inelastic damage analysis of RCMRFs using pushover method. IJST Trans. Civil Eng. 37 (2), 345–352

[27]. Heo, Y., Kunnath, S.K., (2013). Damage-based seismic performance evaluation of reinforced concrete frames. Int. J. Concr. Struct. Mater. 7 (3), 175–182.

[28]. Inel M and Ozmen HB (2006). Effects of plastic hinge properties in nonlinear analysis of reinforced concrete buildings. Engineering structures, 28: 1494-1502

[29]. IS 13920:1993 (1993). Ductile-detailing of reinforced concrete structures subjected to seismic forces-code of practice, Bureau of Indian standards, New Delhi.

[30]. IS 1786, 2008. Indian standard for high strength deformed steel bars and wires for concrete reinforcement. Bureau of Indian standards, New Delhi.

[31]. IS 1893, 2002. Indian standard for earthquake resistant design of structures (part 1): general provisions and buildings. Bureau of Indian standards, New Delhi.

[32]. IS 456, 2000. Indian standard plain and reinforced concrete – code of practice (fourth revision). Bureau of Indian standards, New Delhi.

[33]. IS 875:1987 (Part 1) (1987). Indian standard code of practice for design loads (other than earthquake) for buildings and structures-dead loads. Bureau of Indian standards, New Delhi.

[34]. IS 875:1987 (Part 2) (1987). Indian standard code of practice for design loads (other than earthquake) for buildings and structures-imposed loads. Bureau of Indian standards, New Delhi.

[35]. Joe Maffei (2000). Suggested improvements to performance-based seismic guidelines, 12th world conference on earthquake engineering, Auckland, New Zealand, 30 January- 4 February 2000

[36]. Korkmaz KA and Irtem E (2008). Evaluation of previous and current performance-based analysis methods. In: 14th world conference on earthquake engineering, Beijing, China, 12-17 October 2008

[37]. Kumar R CM, Choudhary V, Babunarayan KS, Reddy V (2014). Moment curvature characteristics for structural elements of RC building.

Journal on Today's Ideas- Tomorrow's Technologies, Vol. 2, No.1, pp. 1-17

[38]. Lagaros ND and Fagiadakis (2011). Evaluation of ASCE 41, ATC 40 and N2 static pushover methods based on optimally designed buildings. Soil dynamic and earthquake engineering, 31: 77-90

[39]. Moehle JP (1992). Displacement-based design of RC structures. Earthquake engineering, 10th world conference, Balkema, Rotterdam; pp. 4299–302

[40]. Mondal A, Siddhartha G, Reddy GR (2013). Performance-based evaluation of the response reduction factor for ductile RC frames. Engineering Structures 56, pp.1808-1819

[41]. Mondal, A., Ghosh, S., Reddy, G.R., (2013). Performance-based evaluation of the response reduction factor for ductile RC frames. Eng. Struct. 56, 1808–1819.

[42]. Muljati I, Fransiscus A, and Willyanto K (2015). Performance of force-based design versus displacement-based design in predicting seismic demands of regular concrete special moment resisting frames. Procedia engineering 125: 1050-1056

[43]. Murthy CVR, Goswami R, Vijaynarayan AR, Mehta VV (2012). Some concepts in Earthquake behavior of buildings, Gujarat State Disaster Management Authority, Government of Gujarat.

[44]. Park R and Paulay T (1975). Reinforced concrete structures. John Wiley and sons, New York. 496-512.

[45]. Priestley MJN, Calvi GM, and Kowalsky MJ (2007). Displacement-based seismic design of structures. IUSS Press, Pravia, Italy.

[46]. Priestly, M., (1997). Displacement-based Seismic Assessment of Reinforced Concrete buildings. J. Earthquake. Eng. 1 (1), 157–192.

[47]. Putul Haldar and Yogendra Singh (2009) Seismic performance and vulnerability of Indian code-designed RC frame buildings. ISET Journal of Earthquake Technology, Paper No. 502, Vol. 46, No. 1, pp. 29–45

[48]. Rakesh K. Goel, Charles Chadwell (2008). Evaluation of ASCE-41 Nonlinear Static Procedure using Recorded Motions of Reinforced-Concrete Buildings. Structures Congress.

[49]. SEAOC vision 2000 (1995). Performance based seismic engineering of buildings, vol. I and vol. II: conceptual framework, Structural Engineering Association of California.

[50]. Sujin S. George, Valsson Varghese (2012). General Concepts of Capacity Based Design, International Journal of Innovative Technology and Exploring Engineering (IJITEE) ISSN: Volume 1, 2278-3075

[51]. Sunitha P, Goswami R, and Murthy CVR (2016). Idealised bilinear moment-curvature curves of RC sections for pushover analysis of RC frame buildings. The Indian concrete journal, 90 (4): 43-54

[52]. Wang JF, Lin CC, Yen SM, (2007). A story damage index of seismically-excited buildings based on modal frequency and mode shape. Eng Struct, 29(9):2143–57.

[53]. Whittaker A, Hart G, Rojahn C (1999). Seismic response modification factors, Journal of Structural Engineering, 125 (4): pp 438-444

[54]. Wilson EL and Habibullah A (2000) SAP 2000/NL-Push software. Computer and structures, Inc. Berkeley (CA), Version 17

[55]. Wilson, E.L., Habibullah, A., (2000). SAP 2000/NL-push version 20 software, computer and structures, Inc, Berkeley, CA, USA

[56]. Wiratman Wangsadinata Ir (1999). Capacity design, a concept to ensure seismic resistance of building structures. First National Conference on Earthquake Engineering, Institute Teknologs Bandung, November 4-5

[57]. Zameeruddin M and Sangle KK (2016). Review on recent developments in the performance- based seismic design of reinforced concrete structures. Structures, 6: 119-133

[58]. Zameeruddin M and Sangle KK (2017a). Seismic damage assessment of reinforced concrete structures using nonlinear static analysis. KSCE journal of civil engineering, 21 (4): 1319- 1330

[59]. Zameeruddin M and Sangle KK (2017b). Seismic performance evaluation of reinforced concrete frames subjected to seismic loads. Journal of the Institution of Engineers (India): Series A, 98 (12): 177-183

[60]. Zameeruddin M and Sangle KK (2017c). Energy-based damage assessment of RCMRFS using pushover. Asian journal of civil engineering (BHRC) 18 (7): 1077-1093

[61]. Zameeruddin, M., Sangle, K.K., (2016). Review on recent developments in performance-based seismic design of reinforced concrete structures. Structures 6, 119–133.

[62]. Zameeruddin, M., Sangle, K.K., (2017a). Seismic damage assessment of reinforced concrete structure using nonlinear static analyses. KSCE J. Civil Eng. 21 (4), 1319–1330.

[63]. Zameeruddin, M., Sangle, K.K., (2017b). Seismic performance evaluation of reinforced concrete frames subjected to seismic loads. J. Inst. Eng. India Ser. A 98, 177–183.

[64]. Zameeruddin, M., Sangle, K.K., (2017c). Energy-based damage assessment of RCMRFs using pushover. Asian J. Civil Eng. (BHRC) 18 (7), 1077–1093.

[65]. Zameeruddin, M., Sangle, K.K., (2021a). Performance-based seismic assessment of reinforced concrete frames. J. King Saud Univ. Eng. Sci. 33, 153–165.

[66]. Zameeruddin, M., Sangle, K.K., (2021b). Damage assessment of reinforced concrete moment resisting frames using performance-based seismic evaluation procedure. J. King Saud Univ. Eng. Sci. 33, 227-239.

[67]. Zou XK, Chan CM (2005) Optimal seismic performance-based design of reinforced concrete building using nonlinear pushover analysis. Engineering Structures; 27(8): 1289–302

Appendix-I
Collapse Mechanism

Table 7.1: *Mechanism for the Collapse of Example MRFs subjected to Push-1 Load case*

Operational Level (OP)								
Series 1			Series 2			Series 3		
Storey Height	Damage Value (Strength)	Damage Value (Stiffness)	Storey Height	Damage Value (Strength)	Damage Value (Stiffness)	Storey Height	Damage Value (Strength)	Damage Value (Stiffness)
0	0	0	0	0	0	0	0	0
3.000	0.343	0.092	3.000	0.789	0.103	3.000	0.567	0.020
9.000	0.591	0.258	9.000	0.766	0.192	9.000	0.736	0.210
15.000	0.553	0.353	15.000	0.747	0.291	15.000	0.747	0.330
21.000	0.681	0.385	21.000	0.724	0.373	21.000	0.727	0.384
27.000	0.774	0.459	27.000	0.663	0.381	27.000	0.577	0.298
33.000	0.713	0.458	33.000	0.603	0.358	33.000	0.613	0.395
39.000	0.790	0.335	39.000	0.611	0.341	39.000	0.610	0.382
45.000	0.795	0.426	45.000	0.620	0.345	45.000	0.526	0.331

Life Safety (LS)								
Series 1			Series 2			Series 3		
Storey Height	Damage Value (Strength)	Damage Value (Stiffness)	Storey Height	Damage Value (Strength)	Damage Value (Stiffness)	Storey Height	Damage Value (Strength)	Damage Value (Stiffness)
0	0	0	0	0	0	0	0	0

3.000	0.991	0.934	3.000	0.926	0.789	3.000	0.926	0.788
9.000	0.916	0.793	9.000	0.870	0.693	9.000	0.859	0.708
15.000	0.910	0.687	15.000	0.840	0.674	15.000	0.871	0.721
21.000	0.865	0.715	21.000	0.854	0.714	21.000	0.865	0.685
27.000	0.864	0.664	27.000	0.900	0.846	27.000	0.787	0.689
33.000	0.856	0.648	33.000	0.928	0.532	33.000	0.812	0.607
39.000	0.950	0.658	39.000	0.808	0.596	39.000	0.819	0.628
45.000	0.992	0.653	45.000	0.811	0.610	45.000	0.876	0.466

Collapse (C)								
Series 1			Series 2			Series 3		
Storey Height	Damage Value (Strength)	Damage Value (Stiffness)	Storey Height	Damage Value (Strength)	Damage Value (Stiffness)	Storey Height	Damage Value (Strength)	Damage Value (Stiffness)
0	0	0	0	0	0	0	0	0
3.000	0.991	0.934	3.000	0.994	0.931	3.000	0.995	0.930
9.000	0.901	0.921	9.000	1.000	0.882	9.000	1.000	0.881
15.000	1.000	0.885	15.000	1.000	0.866	15.000	1.000	0.787
21.000	0.996	0.856	21.000	1.000	0.858	21.000	1.000	0.857
27.000	0.998	0.836	27.000	1.000	0.846	27.000	0.994	0.882
33.000	1.000	0.818	33.000	1.075	0.810	33.000	0.995	0.827
39.000	0.909	0.878	39.000	0.995	0.810	39.000	0.994	0.815
45.000	0.927	0.862	45.000	0.993	0.806	45.000	1.000	0.787

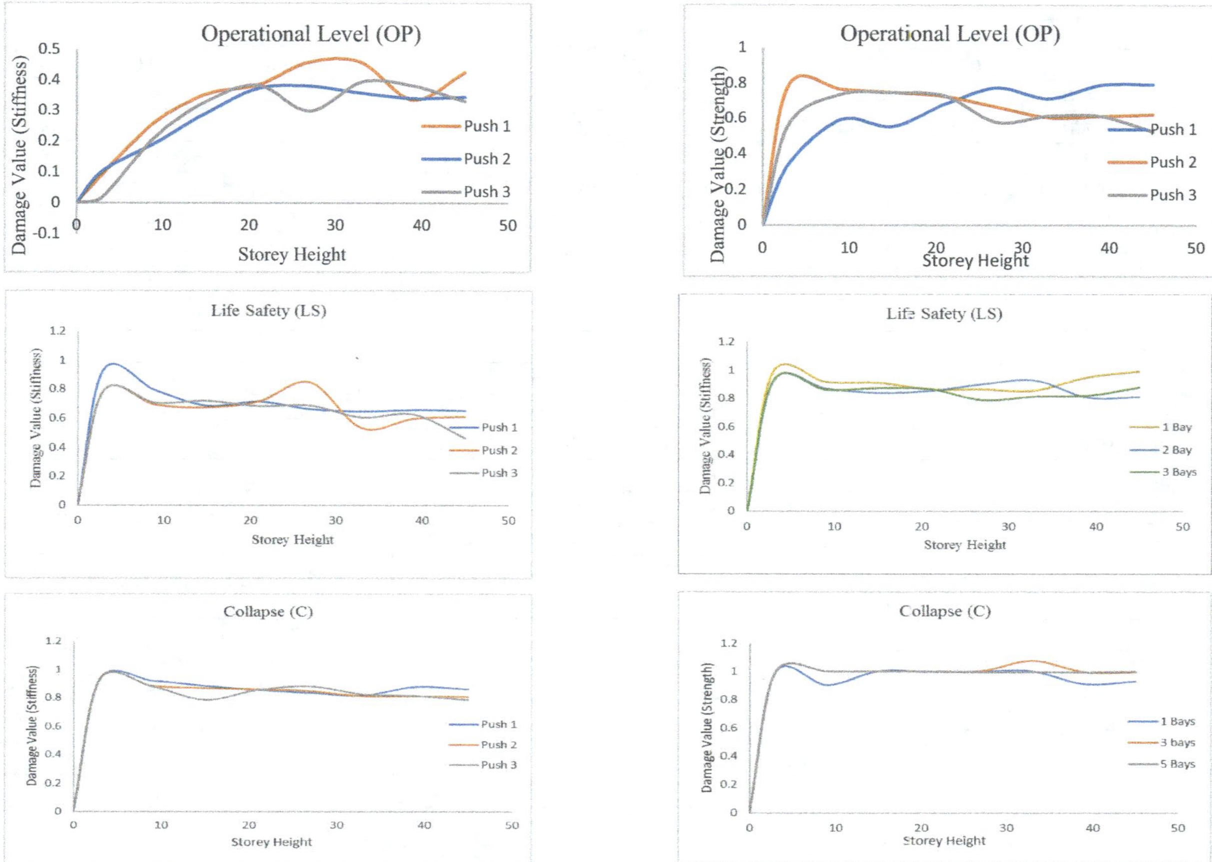

Figure 7.1: ***Graph Showing damage variations in Series-I, Series-II & Series-III MRFs with three storeys subjected to different Push-I Load***

Table 7.2: ***Mechanism for the Collapse of Example MRFs subjected to Push-2 Load case***

Operational Level (OP)								
Series 1			Series 2			Series 3		
Storey Height	Damage Value (Strength)	Damage Value (Stiffness)	Storey Height	Damage Value (Strength)	Damage Value (Stiffness)	Storey Height	Damage Value (Strength)	Damage Value (Stiffness)
0	0	0	0	0	0	0	0	0
3.000	0.243	0.092	3.000	0.263	0.092	3.000	0.253	0.082
9.000	0.139	0.358	9.000	0.129	0.258	9.000	0.239	0.458
15.000	0.186	0.353	15.000	0.176	0.253	15.000	0.286	0.253
21.000	0.165	0.385	21.000	0.169	0.285	21.000	0.265	0.285
27.000	0.120	0.379	27.000	0.130	0.279	27.000	0.220	0.379
33.000	0.142	0.326	33.000	0.152	0.303	33.000	0.242	0.326
39.000	0.113	0.374	39.000	0.191	0.314	39.000	0.313	0.374
45.000	0.160	0.326	45.000	0.150	0.316	45.000	0.260	0.356

Life Safety (LS)								
Series 1			Series 2			Series 3		
Storey Height	Damage Value (Strength)	Damage Value (Stiffness)	Storey Height	Damage Value (Strength)	Damage Value (Stiffness)	Storey Height	Damage Value (Strength)	Damage Value (Stiffness)
0	0	0	0	0	0	0	0	0
3.000	0.919	0.773	3.000	0.919	0.573	3.000	0.919	0.773
9.000	1.052	0.772	9.000	1.152	0.972	9.000	1.152	0.972
15.000	0.690	0.686	15.000	0.690	0.586	15.000	0.790	0.586

21.000	0.692	0.481	21.000	0.792	0.581	21.000	0.892	0.681
27.000	0.720	0.480	27.000	0.720	0.490	27.000	0.720	0.590
33.000	0.960	0.631	33.000	0.960	0.431	33.000	0.996	0.731
39.000	0.902	0.550	39.000	0.992	0.750	39.000	0.992	0.850
45.000	0.935	0.515	45.000	0.935	0.615	45.000	0.835	0.815

Collapse (C)								
Series 1			Series 2			Series 3		
Storey Height	Damage Value (Strength)	Damage Value (Stiffness)	Storey Height	Damage Value (Strength)	Damage Value (Stiffness)	Storey Height	Damage Value (Strength)	Damage Value (Stiffness)
0	0	0	0	0	0	0	0	0
3.000	0.991	0.932	3.000	1.091	0.732	3.000	1.191	0.732
9.000	1.160	0.925	9.000	1.160	0.725	9.000	1.360	0.825
15.000	0.763	0.890	15.000	0.763	0.690	15.000	0.563	0.690
21.000	0.971	0.819	21.000	1.092	0.682	21.000	1.092	0.719
27.000	0.960	0.801	27.000	0.960	0.701	27.000	0.760	0.801
33.000	0.639	0.704	33.000	0.639	0.804	33.000	0.639	0.504
39.000	0.814	0.853	39.000	0.914	0.553	39.000	0.814	0.753
45.000	0.596	0.838	45.000	0.796	0.538	45.000	0.696	0.638

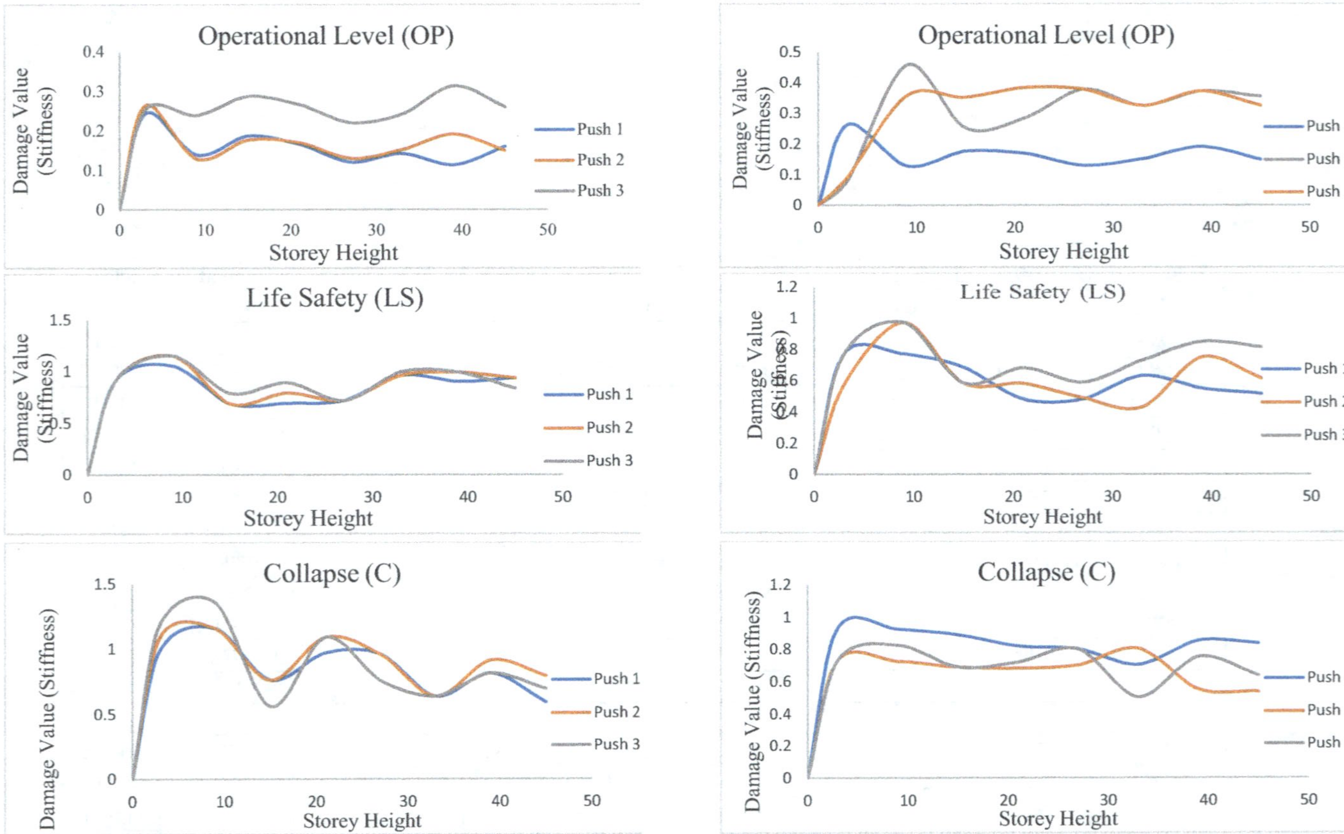

Figure 7.2: ***Graph Showing damage variations in Series-I, Series-II & Series-III MRFs with three storeys subjected to different Push-***

Table 7.3: ***Mechanism for the Collapse of Example MRFs subjected to Push-3 Load case***

Operational Level (OP)										
Series 1				Series 2				Series 3		
Storey Height	Damage Value (Strength)	Damage Value (Stiffness)		Storey Height	Damage Value (Strength)	Damage Value (Stiffness)		Storey Height	Damage Value (Strength)	Damage Value (Stiffness)
0	0	0		0	0	0		0	0	0
3.000	0.000	0.230		3.000	0.250	0.430		3.000	0.350	0.330
9.000	0.252	0.450		9.000	0.352	0.450		9.000	0.352	0.250
15.000	0.151	0.950		15.000	0.251	0.750		15.000	0.151	0.750
21.000	0.121	0.990		21.000	0.221	0.990		21.000	0.121	0.970
27.000	0.202	0.750		27.000	0.302	0.950		27.000	0.202	0.850
33.000	0.247	0.650		33.000	0.447	0.650		33.000	0.147	0.450
39.000	0.212	0.450		39.000	0.312	0.450		39.000	0.212	0.450
45.000	0.119	0.350		45.000	0.119	0.550		45.000	0.129	0.450

Life Safety (LS)										
Series 1				Series 2				Series 3		
Storey Height	Damage Value (Strength)	Damage Value (Stiffness)		Storey Height	Damage Value (Strength)	Damage Value (Stiffness)		Storey Height	Damage Value (Strength)	Damage Value (Stiffness)
0	0	0		0	0	0		0	0	0
3.000	0.597	0.776		3.000	0.497	0.976		3.000	0.797	0.676
9.000	0.838	0.637		9.000	0.738	0.837		9.000	0.838	0.737

15.000	0.816	1.000	15.000	0.716	1.200	15.000	1.059	0.800
21.000	0.777	0.850	21.000	0.777	0.950	21.000	0.977	0.750
27.000	0.748	0.647	27.000	0.748	0.547	27.000	0.848	0.547
33.000	0.679	0.419	33.000	0.579	0.519	33.000	0.879	0.619
39.000	0.212	0.350	39.000	0.412	0.150	39.000	0.812	0.450
45.000	0.263	0.387	45.000	0.463	0.587	45.000	0.463	0.587

Collapse (C)								
Series 1			Series 2			Series 3		
Storey Height	Damage Value (Strength)	Damage Value (Stiffness)	Storey Height	Damage Value (Strength)	Damage Value (Stiffness)	Storey Height	Damage Value (Strength)	Damage Value (Stiffness)
0	0	0	0	0	0	0	0	0
3.000	0.666	0.929	3.000	0.866	1.129	3.000	0.766	0.929
9.000	0.998	0.872	9.000	0.998	1.372	9.000	0.898	0.772
15.000	0.987	1.000	15.000	0.987	1.000	15.000	0.787	1.300
21.000	0.802	0.518	21.000	0.502	0.918	21.000	0.602	0.518
27.000	0.991	0.875	27.000	0.991	0.975	27.000	0.891	0.575
33.000	0.951	0.763	33.000	0.951	0.863	33.000	0.851	0.663
39.000	0.706	0.398	39.000	0.806	0.798	39.000	0.706	0.798
45.000	0.959	0.758	45.000	0.659	0.858	45.000	0.670	0.858

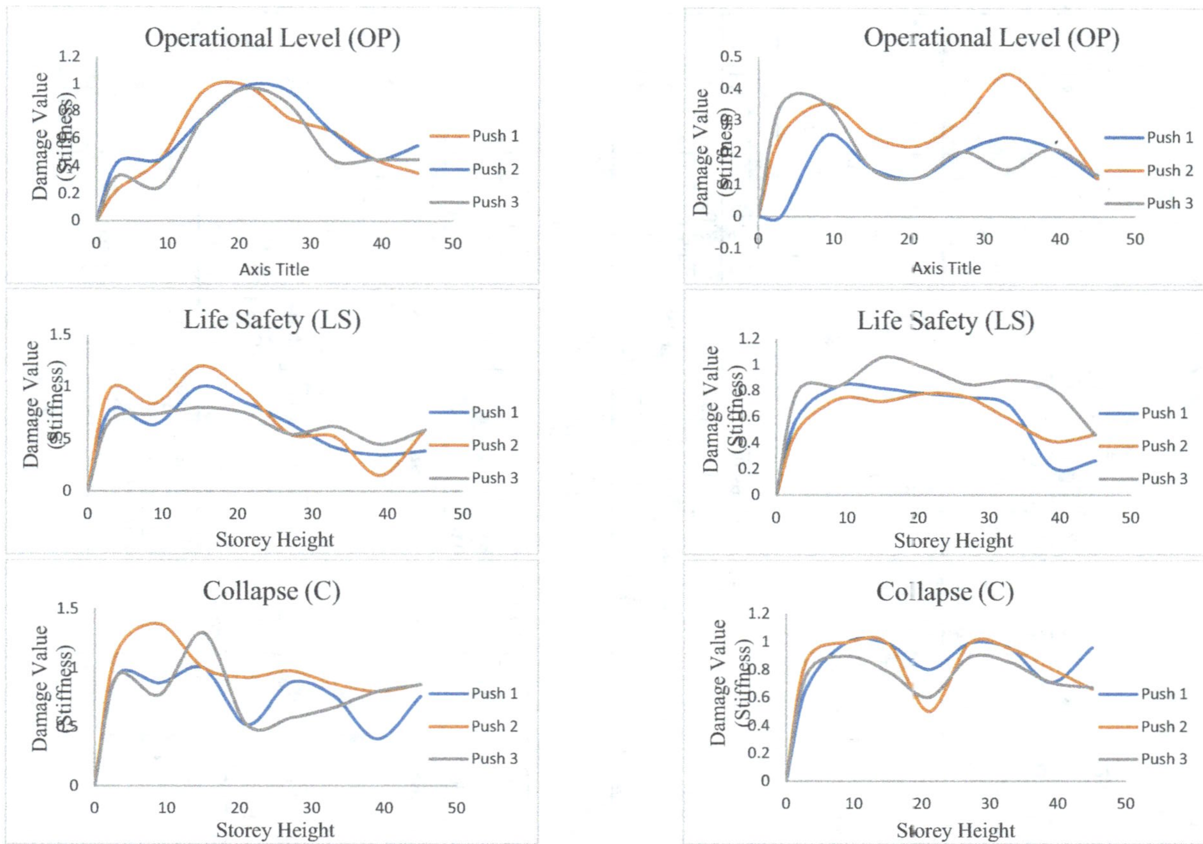

Figure 7.3: ***Graph Showing damage variations in Series-I, Series-II & Series-III MRFs with three storeys subjected to different Push-3 Load***

Table 7.4: ***Comparison of drift limits for various Push-1, Push-2 & Push-3 load cases with Performance Levels***

PUSH 1

Performance Level	Drift Limit in % (Permissible)	**Series 1**							
		S1B1	S3B1	S5B1	S7B1	S9B1	S11B1	S13B3	S15B1
OP	0.2	0.00	0.01	0.01	0.01	0.05	0.67	0.04	0.05
LS	1.5	0.07	0.22	0.34	0.32	0.46	1.36	0.65	0.73
C	>2.5	0.23	0.69	1.15	1.46	1.49	3.91	2.53	2.39
Performance Level	Drift Limit in % (Permissible)	**Series 2**							
		S1B3	S3B3	S5B3	S7B3	S9B3	S11B3	S13B3	S15B3
OP	0.2	0.00	0.01	0.02	0.02	0.02	0.00	0.04	0.07
LS	1.5	0.06	0.19	0.30	0.36	0.49	0.05	0.77	0.94
C	>2.5	0.23	0.68	1.18	1.63	2.08	1.36	2.93	3.53
Performance Level	Drift Limit in % (Permissible)	**Series 3**							
		S1B5	S3B5	S5B5	S7B5	S9B5	S11B5	S13B3	S15B5
OP	0.2	0.00	0.01	0.01	0.02	0.02	0.04	0.06	0.01
LS	1.5	0.06	0.19	0.28	0.39	0.43	0.53	0.57	0.64
C	>2.5	0.23	0.69	1.13	1.64	1.62	2.35	2.23	2.68

PUSH 2

Performance Level	Drift Limit in % (Permissible)	**Series 1**							
		S1B1	S3B1	S5B1	S7B1	S9B1	S11B1	S13B3	S15B1
OP	0.2	0.00	0.01	0.01	0.01	0.04	0.57	0.03	0.05
LS	1.5	0.07	0.22	0.34	0.35	0.55	1.54	0.75	0.83
C	>2.5	0.35	0.79	1.25	1.56	1.69	3.01	2.65	2.59
Performance Level	Drift Limit in % (Permissible)	**Series 2**							
		S1B3	S3B3	S5B3	S7B3	S9B3	S11B3	S13B3	S15B3
OP	0.2	0.00	0.01	0.02	0.02	0.02	0.00	0.03	0.06
LS	1.5	0.07	0.20	0.35	0.40	0.49	0.05	0.77	0.95
C	>2.5	0.25	0.70	1.28	1.69	2.58	1.46	2.93	3.60
Performance Level	Drift Limit in % (Permissible)	**Series 3**							
		S1B5	S3B5	S5B5	S7B5	S9B5	S11B5	S13B3	S15B5
OP	0.2	0.00	0.12	0.15	0.02	0.04	0.04	0.06	0.01
LS	1.5	0.02	0.29	0.30	0.39	0.45	0.54	0.60	0.65
C	>2.5	0.25	0.70	1.17	1.69	1.64	2.39	2.29	2.60

PUSH 3

Performance Level	Drift Limit in % (Permissible)	Series 1							
		S1B1	S3B1	S5B1	S7B1	S9B1	S11B1	S13B3	S15B1
OP	0.2	0.00	0.01	0.01	0.01	0.05	0.60	0.04	0.05
LS	1.5	0.08	0.32	0.44	0.45	0.65	1.84	0.85	0.93
C	>2.5	0.45	0.89	1.35	1.66	1.79	3.11	2.75	2.69
Performance Level	Drift Limit in % (Permissible)	Series 2							
		S1B3	S3B3	S5B3	S7B3	S9B3	S11B3	S13B3	S15B3
OP	0.2	0.01	0.07	0.01	0.03	0.03	0.00	0.04	0.07
LS	1.5	0.05	0.09	0.55	0.50	0.59	0.05	0.87	0.95
C	>2.5	0.35	0.80	1.38	1.79	2.78	1.66	2.93	3.80
Performance Level	Drift Limit in % (Permissible)	Series 3							
		S1B5	S3B5	S5B5	S7B5	S9B5	S11B5	S13B3	S15B5
OP	0.2	0.00	0.22	0.25	0.03	0.05	0.05	0.07	0.05
LS	1.5	0.02	0.29	0.30	0.59	0.55	0.56	0.70	0.85
C	>2.5	0.27	0.70	1.17	1.69	1.84	2.49	2.39	2.76

Generis

PUBLISHING